GIS Solutions in Natural Resource Management: Balancing the Technical-Political Equation

Stan Morain, Editor

GIS Solutions in Natural Resource Management: Balancing the Technical-Political Equation

Stan Morain, Ed.

Published by:

OnWord Press
2530 Camino Entrada
Santa Fe, NM 87505-4835 USA
Carol Leyba, Publisher
David Talbott, Acquistions Director
Barbara Kohl, Associate Editor
Daril Bentley, Senior Editor
Cynthia Welch, Production Manager
Elizabeth Bennie, Marketing Director
Lauri Hogan, Marketing Services Manager
Lynne Egensteiner, Cover designer, Illustrator
Nan Badgett, Indexer

First edition, 1999

SAN 694-0269

10 9 8 7 6 5 4 3 2 1

Printed in the United States of America

Library of Congress Cataloging-in-Publication Data
GIS solutions in natural resource management / Stan Morain, ed.

    p.   cm.

    Includes bibliographical references and index.

    ISBN 1-56690-146-4

    1. Geographic information systems. 2. Natural resources--Management. I. Morain, Stanley A.
G60.212.G58   1998
910' .285--dc21

                                      97-45605
                                        CIP

Trademarks

Warning and Disclaimer

Contents

Introduction xiii

Section I
Approaches to Natural Resource Management   1

Chapter 1: Development of the Spatial Domain in Resource Management 5
Historical Context 6
Examples of GIS Management Models 10
    Descriptive 10
    Predictive 11
    Prescriptive 11
Using GIS to Create Models 12
    Data Exploration and Analysis 12
    Algorithm Specification 12
    Accuracy Assessment 13
    Future Directions 14
References 15

Chapter 2: Spatial Pattern Analysis Techniques 17
Landscape Ecology 18
Point versus Patch Analysis Techniques 19
Point Data Analysis Techniques 22
    Nearest-neighbor Methods 22
    Join-count Analysis 22
Lacunarity Analysis 26
References 34

Section II
GIS Requirements
for Natural Resource Management   41

Chapter 3 : Multi-scale Framework Data 45

   NSDI Strategy 47

   Framework Overview 48

      Goals of Framework Infrastructure 48

      Composition of Framework 48

      Implementation 51

   Data Themes 52

      Geodetic Control 52

      Digital Orthoimagery 52

      Elevation Data 53

      Transportation 54

      Hydrography 54

      Governmental Units 55

      Cadastral 55

   Framework Issues 56

   Technical Issues 56

      Coordinate Data Development 56

      Implement Standard Practices 57

      Horizontal and Vertical Data Integration 57

      Implement Metadata Standards 58

      Feature Based Updating 60

      Archive Data 62

   Administrative Issues 63

      Policies 63

      Theme Expertise 64

      Framework Management 64

      Area Integration 64

      Data Production 66

      Data Distribution 66

   Implementing Framework 67

   Summary 70

   References 70

Harmonizing Framework and Resource Data across Political Boundaries: The Tijuana River Watershed GIS 71

Resource Management Requirement 71

Objectives and Phases 74

Harmonizing Framework Data 76

Hypsography and Derived Themes 76

Hydrography 78

Roads 79

Harmonizing Resource Data 81

Vegetation 81

Soils 83

Geology 87

Geomorphology 87

Land Use 89

Census Enumeration Units 90

Summary 91

Acknowledgments 92

Endnote 92

References 92

A Modular Ground Water Modeling System 95

Resource Management Requirement 95

GWZOOM Concept 96

GWZOOM System 97

Implementation 97

Results 101

Conclusions 103

References 104

Locating the Continental Divide Trail in New Mexico 105

Resource Management Requirement 105

Challenges in New Mexico 105

Enabling Legislation 107

Multiple Interests 107

Government Agencies 107

Continental Divide Trail Society 108

Continental Divide Trail Alliance 108

Status of Continental Divide Trail 108

Status of Trail in New Mexico 111

Issues in New Mexico 111

Solution 112

Database Prototype 113

Conclusions 114

Cartographic Support for Managing Washington State's Aquatic Resources 115

Research Management Requirement 115

Aquatic Ownership Records Program 116

Shellfish Management Program 118

Nearshore Habitat Program 121

Spartina Control Program 124

Conclusions 124

Chapter 4: Multi-scale Resource Data 125

Map Scale 129

Map Error 131

Sources of Error 131

Measuring Error 135

Conclusions 138

References 138

Modeling Conservation Priorities in Veracruz, Mexico 141

Resource Management Requirement 141

Methodology 142

Results 147

References 150

Development of a GIS Hydrography Data Base of Phosphorus Transport Modeling in the Lake Okeechobee Watershed 151

Resource Management Requirement 151

Basin Description 152

Database Development 154

Hydrography Coverage 154

Linkage with Cross-section Data 155

Stream Reach Numbering 157

Results and Discussions 160

Summary 162

Acknowledgments 162

References 162

Chapter 5: Indicators of Resources and Landscapes 165

Composition Metrics 167

Pattern Metrics 169

Dominance and Diversity Indices 170

Connectivity Indices 171

Fragmentation Indices 172

Shape Metrics 173

Scale Metrics 175

Discussion 177

Conclusion 178

Acknowledgment 178

References 178

Global Vegetation Production and Human Activity 183

Resource Management Requirement 183

Methodology 184

Results 185

Other Factors 189

Conclusions 190

Acknowledgments 191

References 191

**Using Population Data to Address the Human Dimensions
of Environmental Change** 193

Resource Management Requirement 193

Methods 196

Results and Discussion 198

Acknowledgments 204

References 204

Section III
Translating Applications to Social Significance   207

Chapter 6: Multi-scale Economic and Demographic Data 209

Remote Sensing of Urban/Suburban Attributes 210

Computing Environments for Handling Remotely Sensed Data 211

Experiments with Spatial Resolution 214

Neighborhood Change Detection 219

Conclusions 231

References 233

**Coping Strategies in the Sahel and Horn of Africa: A Conceptual Model
Based on Cultural Behavior and Satellite Sensor Data** 235

Resource Management Requirement 235

Social Science Approaches 235

Earth System Science Approaches 236

Environmental Limits of the Study Area 238

Remotely Sensed Measurements of Primary Production 242

Human Responses to Primary Production and Interannual Variability . 243

Conclusions 248

References 249

**Landscape Characterization through Remote Sensing, GIS,
and Population Surveys** 251

Resource Management Requirement 251

Study Area 252

Data and Methodology 255

Land Cover Changes and GIS Development 255

Population Survey 256

Village Boundaries 256

Topographic Surfaces 259

Pattern Metrics 260

Geographic Accessibility and Competition 261

Results and Discussion 262

Acknowledgments 264

References 265

**Societal Dimensions of Ecosystem Management
in the South Florida Everglades**. 267

   Resource Management Requirement 267

   Conceptualizing and Implementing Ecosystem Management 268

   Methodology 269

      Implementing the GIS 271

      Dependent and Independent Variables 275

   Results and Discussion 277

   Conclusions 280

   Acknowledgments 280

   References 280

People and Place: Dasymetric Mapping Using ARC/INFO. 283

   Resource Management Requirement 283

   Methods 285

   Results 288

   Conclusions 290

   Acknowledgments 291

Chapter 7: Modeling 293

   Paradigm Shift 294

   Models 296

      Good Models 297

      Types of Models 299

   Dynamic Spatial Modeling 304

      Dynamic Models 306

      Anatomy of a Dynamic Spatial Model 308

      Computational Considerations 314

   WWW Resources 316

   References 317

**Analyzing Phosphorus Loads and Transport
in the Lake Okeechobee Watershed** 325

   Resource Management Requirement 325

   Driving Force 326

   Data and Methodology 327

      System Design 327

GIS-PLAT Databases 331
GIS-PLAT User Interface 334
Summary 337
Acknowledgments 337
References 337

Global Commons Risk Assessment 339
Resource Management Requirement 339
Results 342
Conclusion 344
Disclaimer 344
References 344

Planning Emergency Response at a Federal Laboratory 347
Resource Management Requirement 347
Methodology 348
Results 350
References 350

Contributors 351

Index 361

Introduction

GIS Solutions in Natural Resource Management owes its concept to two major conferences held in late 1996 by the Renewable Natural Resources Foundation (RNRF) and the National Academy of Sciences/National Research Council (NAS/NRC). Both conferences were aimed at advancing scientific understanding of the roles spatial and spectral technologies could or should play in better managing natural resources for sustainable purposes. RNRF conducted its Congress on "Applications of Geographic Information Systems to the Sustainability of Renewable Natural Resources" at Jackson Lake Lodge in September. NAS/NRC's Committee on the Human Dimensions of Global Change hosted its meeting on "People and Pixels" in Washington, D.C. in November. Both have issued reports that would serve as further reading.

The 20 member organizations represented at the RNRF Congress came from professional societies in all sectors of natural resource concerns, from wildlife ecologists to landscape architects, from plant physiologists to anthropologists, from range managers to civil engineers, and from soil scientists to meteorologists. Considering the diversity of these scientific and professional interests, and considering that delegates came from all sectors of society, it is significant that GIS and sustainability were selected as the theme and only topic of the Congress. It is equally significant that the delegates approached this topic as a series of issues that seem to hinder adoption of GIS among practicing professionals. People and Pixels, on the other hand, was attended mainly by federal government professionals and academics who focused their dialog on the socioeconomic intricacies surrounding human-impacted natural systems. Their concerns attempted to link local and regional demographics with economic activity and resulting indicators of change in affected landscapes. In both gatherings, it was well known to participants that spatial and spectral analytical technologies have burst into workplaces at all levels of government, private, and nonprofit natural resource/environmental organizations. The technologies are compelling in their application, and as a result of affordable high speed computing, are quickly becoming ubiquitous. Common concerns include making sense of trends and developments, understanding the impact of next generation capabilities on the information highway, and ensuring against runaway applications that cost rather than save natural resources.

Member organizations of the Renewable Natural Resources Foundation

American Anthropological Association
American Congress on Surveying and Mapping
American Fisheries Society
American Geophysical Union
American Meteorological Society
American Society for Horticultural Science
American Society for Photogrammetry and Remote Sensing
American Society of Agronomy
American Society of Civil Engineers
American Society of Landscape Architects
American Society of Plant Physiologists
American Water Resources Association
Association of American Geographers
Society for Range Management
Society of Wood Science and Technology
Soil and Water Conservation Society
The Ecological Society of America
The Humane Society of the United States
The Nature Conservancy
The Wildlife Society

This book is an amalgam of ideas generated by RNRF and NAS/NRC. The aim within Sections I and II is to provide a progression of technical developments and requirements for current resource management applications. Section III goes beyond these traditional approaches and their requirements to translate GIS resource technology into social and economic terms. Each chapter is written in a tutorial style to convey its central ideas to professional resource managers and college level students in nonrigorous terms. With the exception of Chapters 1 and 2, all chapters are accompanied by case studies to highlight chapter content.

In Section I, approaches to natural resource management are presented in a progressive fashion from the non-spatial to the spatial domain. Throughout all but the last decade of the twentieth century natural resource studies were approached in autecological or synecological fashion that produced facts about resources and how they are connected into systems. In the absence of tools more robust than the topographic maps on which sample or specimen sites were recorded, little consideration was, or could be, given to the spatial domain in which the "facts" operated. Between the early to mid-1960s and today, the inauguration of statistical methods advanced understanding by allowing predictions of natural resources to be made over broad areas using discontinuous data from scattered field plots and transects. Many natural resource management strategies active today are based on these techniques, and there is a wealth of archival data that could, if properly documented, be migrated to modern and next generation GISs. The discussion also focuses specifically on the evolution of GIS, technically and scientifically, in meeting resource managers' needs and, at the same time, how the technology serves society at large through what has come to be referred to as "participatory democracy."

Section II addresses GIS requirements for natural resource management. Chapter 3 focuses on multi-scale framework data. These are the data themes recognized by the Federal

Geographic Data Committee (FGDC) as most often required for attribution in resource applications. Resource managers are themselves frequently involved in collecting and developing these framework data at project and local level scales, so it is important that they understand how their data are integrated horizontally and vertically between themes, *and* the need to document these data in standardized metadata protocols.

Chapter 4 focuses on attributing framework data with resource information. For most resource managers, this is the heart of the matter because their funding mandates and responsibilities are geared to resources. In the new paradigm, however, trained resource managers must understand the spatial context of their measurements and data collection strategies in order for others to use their results with confidence. For some applications, it may be possible to migrate archival (legacy) data to the GIS environment; for most others, such data sets may not be usable in the new technology because of low or ambiguous spatial quality. This situation poses a dilemma for managers. To the conservative mind, it is difficult to set aside early data about resources because their value as retrospective attributes for time series analyses, and for visualizing historical trends, is obvious. On the other hand, most modern spatial modeling for resource applications requires that these data have known locational attributes and a documented pedigree. Moreover, it is important that *all* participants in resource management, including academic researchers and local government project personnel, document their data with metadata.

Chapter 5 identifies indicators of environmental and resource conditions and their incorporation into GIS-based resource applications. Attributing a framework with resource data for a project area or jurisdiction is among the early steps in developing a GIS that produces information. A significant later step is knowing what to look for in the data that help interpret landscape changes. Spectral and spatial domains combine to characterize the condition of resources over time. By interpreting these changes and conditions, it is often possible to translate evolving landscape patterns into cultural, socio-economic, or political impacts that link human activity to resource sustainability. How to effect this translation was, after all, a major theme of the People and Pixels conference. It is also a major theme of the U.S. Environmental Protection Agency's Border 21 program along the United States/Mexico North American Free Trade Agreement (NAFTA) region, and is the core idea behind the conceptual model shown in the following figure for an integrated scale-up and scale-down model for Earth system assessments.

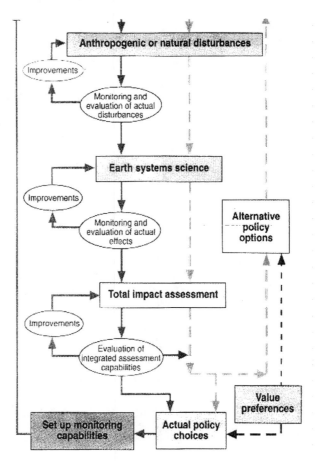

Conceptual model for making incremental improvements in resource management policies. [From Science 269 (1996), 338.]

The above schematic shows human and natural disturbances in Earth systems, and inserts scale-up and scale-
down policy mechanisms to achieve incremental improvements in resource management. The assumption, of course,
is that data about resource conditions are being received on
a regular schedule and that these can be inserted directly
into spatial structures for comparative analyses. Monitoring
actual effects and disturbances implies an understanding of
the indicators of change, and we can safely assume that
many of these indicators have spectral and spatial dimensions.

Section III provides a look at how technical capabilities in
resource management can be translated into social terms.
Because resource management questions are posed most
often in the context of human use of landscapes, Chapter 6
describes multi-scale economic and demographic data. To
manage resources in a sustainable way, GISs must include a
variety of economic, demographic, and cultural attributes
that are themselves highly complex interactions that in general have nothing to do with the resources being impacted.
These are the accumulated actions of individuals who are
not consciously behaving in the interests of the resources
they collectively impact.

Chapter 7 provides a review of spatial and spectral modeling approaches. The traditional approaches have evolved
from earlier statistical approaches that have been improved
for multi-scale applications that range from local to global
areas. These models attempt to characterize resource conditions and to predict future conditions within the constraints
of the model. The aim of managing toward *desired future
conditions* has been widely adopted either explicitly or
implicitly by many federal and state resource agencies. The
computational technology upon which most of these management strategies are based, however, has progressed so
rapidly in context of the Internet that new approaches and
modeling goals are being developed.

One of the more rapidly changing goals is, in fact, a retreat from management for desired future conditions, at least in the strictest sense of humans managing natural landscapes as though they were enclosed environments. The newer approach aims toward management that recognizes the self-adaptive (genetic) foundations of resources and systems to both natural and human-caused changes; and which also model the impacts of humans as a part of those systems. This is the crux of the previous schematic. It takes into account both types of change in a series of steps that compares predicted versus actual changes in order to make incremental improvements. The ability to design such complex models is based more and more on an ability to retrieve vast amounts of data from an expanding variety of sources available on the Internet, and to incorporate those data sets into complex GISs.

Section I

Approaches to Natural Resource Management

Natural resources are derived from finely tuned biological systems, called *ecosystems*. In this one word is implied all the Earth system processes and cycles that culminate in the splendor of life around us. Ecosystems in the traditional sense are complex, self-adaptive units that evolve and change in concert with external biogeochemical forces. Intuitively, to manage natural resources should require managing the ecosystems that provide them, but this has not been the traditional approach. Instead, natural resources management has been a commodity-based process where the commodities are decoupled from the systems that produce them. Management has been a constantly shifting goal that, at alternate swings of the pendulum, has been driven by political, economic, and social agendas, but rarely by science and technology agendas. Over the past 75 years of slow, painstaking research to learn how ecosystems regulate themselves and adjust to atmospheric, geologic, and human activities and impositions, resource managers can now link the science agenda to the human agenda in compelling and credible ways. Thanks also to the Information Age, spectral and spatial technologies have emerged in the past 25 years that enable managers to process massive amounts of data from disparate sources to visualize, model, and predict outcomes from management scenarios.

On Earth today, ecosystems *are*, or *have become* human systems by virtue of the universal impacts of population growth and economic activity. Agricultural, urban, forestry, and aquatic systems are all governed by human resource requirements as much as by natural forces, and therefore require delicate management to achieve sustainable yields. Given today's human population, there are, arguably, no purely natural systems remaining on the planet, including remotely located systems in the sense of being remote from direct, day-to-day human impacts. Antarctic marine and alpine felsenmeer systems are just as much a concern for managers as any more habitable places. So pervasive is the fact of human alteration that the Global Change Research Program (GCRP), International Geosphere/Biosphere Program (IGBP), and Man and Biosphere Program (MAP), among others, are all focused on human relations with, and impacts on, Earth systems. Each of these programs has components that rely on spectral and spatial technologies for management solutions.

What was missing in almost all earlier approaches to resource management was the technology necessary to plan and test management scenarios in near real-time. GIS enables the development of such scenarios that can test static as well as dynamic hypotheses about resource uses, changes, and alternative fates, given modeled inputs and outputs. These capabilities represent a quantum leap forward compared to those available even half a decade ago, and are light years ahead of those available in the 1960s, 1970s, and 1980s, when many of today's practicing professionals were trained.

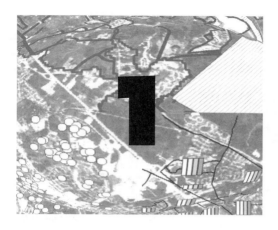

Development of the Spatial Domain in Resource Management

K. Green, Pacific Marine Resources

Geographic information systems are tools used to (1) organize and display spatial information and (2) analyze the spatial impacts of alternative decisions. Their power lies in their ability to manage *spatial* relationships over time. Because natural resource professionals manage landscapes across space and time, GIS helps managers conceive of and implement management alternatives.

While GIS is relatively new, the need to manage natural resources is not. With the closing of the U.S. frontier at the turn of the century came the need for natural resource management systems and procedures. No longer could pioneers march across the nation, exploiting resources as needed and then move on to the next region. When the frontier closed, the need to manage resources efficiently, fairly, and wisely became a growing U.S. mandate. As stated by Dana (1980), "Before the nation could be convinced of the need to conserve resources, those resources had to become valuable; and for that to happen, they had to be recognized as scarce."

Resource management has always required the use of models that (1) describe the resources of interest, (2) predict what will happen to those resources if certain actions are undertaken, and/or (3) prescribe the best course of action given specified goals. Models simplify the world by reducing information to those variables most critically affecting the decisions to be made.

Until the 1980s most of the models were non-spatial. Recently the advent of GIS tools has enabled the development of spatial models. This chapter presents (1) how GIS arose in the context of natural resource management, (2) how it has evolved technically and scientifically to meet the needs of resource managers, and (3) how it serves society at large to address contentious and complex environmental issues. The first section of the chapter explores the history of natural resource decision-making models and how GIS has evolved into an important modeling tool. Next, several real world examples of how GIS is currently used in resource management are presented. The chapter concludes with a discussion of changes required for GIS to remain relevant to continually evolving resource management needs.

Historical Context

Most natural resource models are non-spatial. Examples include *descriptive* models such as timber volume tables; *predictive* models such as wildlife population models and fuel burn intensity equations; and *prescriptive* models such as harvest scheduling programs. For example, a typical forest management model might predict the growth of a tree as a function of the tree crown volume as follows:

$$Growth = 1.136 + 0.059 * Crown\ Volume$$

Late in the nineteenth century, Pinchot and Fernow brought the German concepts of silviculture and sustained yield to the United States. Silviculture is the science of growing forests. It contains models that predict forest growth as a function of inputs over time. Sustained yield is a prescriptive model that relies on the models of silviculture to allocate harvests over time and forest type such that yields from forests are sustained over the long run. Forged in the 1800s of even greater scarcity in Europe of both land and trees, sustained yield philosophies tended to encourage the production of forest material with little regard to other forest resources, the cost required to produce the material, or society's willingness to pay for the resources (Dana 1980).

Models of forest economics that emerged in the 1940s dominated forest and natural resource management and policy analysis until the 1980s. Economics is a set of theories about how humans will and should behave under certain conditions. Economic models can be both descriptive and predictive (how do humans behave), and prescriptive (how should humans behave given specified goals).

Economic models based on forest resources differ from general economic models in the following three ways.

- The periods required to grow forests are considerably longer than most production periods (e.g., manufacturing cycles for automobiles or computers, and growth cycles for cattle or soybeans).

- Trees are both production facility (i.e., factory) and product.

- Many resources derived from forests (such as wildlife habitat and recreational experiences) do not pass through traditional markets and are therefore inadequately priced (Vaux 1953).

Forest economics, and subsequently, natural resource economics grew into very sophisticated fields, resulting in advanced decision making models. Predictive models such as SERTS (Abt forthcoming) forecast future wood fiber prices based on current inventory, growth, and projected harvests. Prescriptive operations research models such as FORPLAN prescribe courses of management action over time given a specified goal and statements of constraints (e.g., maximize the present net worth of a forest given the constraints of reduced cutting in streamside zones and the need to provide the mill with a minimum flow of material over time).

Forest economics primarily focuses on decisions concerning the extensive margin (what acres should be put into the production of forest products) and the intensive margin (how intensively should each type of acre be managed). Both questions require an assessment of current forest inventory and projections of growth and harvest, which in turn requires maps.

Maps are models. They simplify views of the landscape by categorizing features of interest such as forest species groups, habitat condition, road classes, soil types, and so on.

Maps have been an everyday part of American natural resource management since the Broad Arrow Act of 1691 reserved for the British Crown all trees 24 inches or more on land not otherwise granted to a private person. More sophisticated natural resource models have emerged with increasing resource scarcity.

Early uses of maps in natural resource management were limited to assessing the volume of wood fiber available in a forest. In the 1890s, the U.S. Geological Survey created maps of remaining western public domain forestlands under consideration for inclusion in the national forest system. Forests were classified into five classes of thousand board feet per acre. The U.S. Forest Service continued to create maps of forest type across private and public lands well into the 1940s for the assessment of forest resource supplies. The primary purpose of the maps was to provide acres for expanding per acre sample measurements of fiber inventory. However, in the 1950s the cost of creating forest type maps led the Forest Service to rely completely on sampling rather than mapping the landscape. While individual national forests and private landowners continued to map their own properties, few maps were created that portrayed the distribution of natural resources across ownership boundaries.

In 1960 Congress passed the Multiple Use - Sustained Yield Act which codified the importance of both the timber *and* non-timber uses of forests on federal lands. However, management emphasis continued on the sustained production of units such as recreation days, animal unit months, and board feet or cubic feet of wood fiber. Most decision analysis models involved the investigation of unit, cost, and revenue trade-offs of various management alternatives.

The environmental activism of the 1960s and 1970s drove the emphasis of natural resource management away from the production of units and towards the analysis of landscape and location. Many controversies, such as those concerning spotted owl and red cockaded woodpecker habitat, focused on the management of ecosystems and watersheds, regardless of ownership boundaries. Other controversies dealt with the preservation of specific locations such as Redwood National Park. As a result, by the 1980s natural resource management refocused to a combination of sustaining yields *and* ecosystems, requiring mangers to balance the use of ecosystem products while maintaining ecosystem function and aesthetic values. Unidimensional unit based (e.g., jobs per year, board feet per acre) models focused on what could and should be generally removed from the land were no longer sufficient because they did not address spatial issues or the complex interrelationships among natural resources.

Over the last ten years, GIS and remote sensing have emerged as promising tools for natural resource management because they allow for the assessment and modeling of spatial relationships. Geographic information systems are not models, but are computer tools for organizing, accessing, displaying and analyzing spatial information. They are used to create spatial models.

Examples of GIS Management Models

GIS models, like all decision tools, include models that define, predict, and prescribe. Model types and examples are summarized below.

Descriptive

Models that *define* combine inputs to create a new set of output variables based on a set of classification rules. Most GIS models are of this type and include "siting" applications that identify the location of desirable sites as a function of a set of spatial variables. Defining models either specify an identity (e.g., regulations may preclude timber harvesting within a 75' buffer of all class II streams) or specify a functional relationship.

For example, the Insurance Services Office (New York, New York) recently used GIS to define fire hazard in California as a function of fuel, slope, and road access (Green 1998). Another defining GIS was created by the State of Washington to prioritize watersheds for ground assessment based on their susceptibility to cumulative effects from forest harvesting and road building (Bernath et al. 1992). Cumulative effects are changes to the environment caused by the spatial and temporal interaction of natural ecosystem processes resulting from two or more forest practices. The State's hydrologists developed an equation to prioritize forest basins based on GIS coverages of forest type, current land cover, precipitation, rain on snow isohedrals, soil hazard ratings, slope, andromonous fish presence or absence, and the location of fish hatcheries.

Predictive

Models that *predict* behavior specify functions that change input variables based on assumed relationships between the variables. Output maps show predicted changes in input variables and the creation of new variables. For example, the GIS model *FIRE!* (Green et al. 1995) integrates coverages of fuel, crown closure, and digital elevation models with temporal weather, wind, and fuel moisture to predict wildfire behavior across time and space. Output GIS layers include predicted fire perimeters, time of fire arrival, heat, fireline intensity, rate of spread, and flame length.

Prescriptive

Models that *prescribe* access variables that describe and predict plus specified goal functions to prescribe a certain set of actions that should be taken. For example, the harvest scheduling matrix generator FORPLAN submits non-spatial information about the land (e.g., tables summarizing acres by stand type), potential management actions for each type, and the revenue and costs for each action to a linear program. The linear program maximizes or minimizes a stated objective function (e.g., maximize present net worth) subject to constraints (e.g., minimum harvests cannot fall below 100 million board feet per year). Unfortunately, FORPLAN linear programming results are non-spatial. While the results will prescribe the number of acres to receive specific treatments over time, they cannot indicate what specific acres are to be treated. As a result, FORPLAN results may often not be implementable on the ground in the current era of harvest adjacency constraints and concerns over habitat fragmentation. To address this problem, Boise Cascade Corporation developed the Spatial Feasibility Test to desegregate FORPLAN results across the landscape, subject to spatial constraints (Landrum et al. 1996).

Using GIS to Create Models

The above examples show how GIS is used to specify different types of models. GIS can also be used to analyze data for the creation of a model. Building a model requires (1) data exploration and analysis, (2) algorithm specification, and (3) accuracy assessment. GIS can be useful in all three steps.

Data Exploration and Analysis

Data exploration aids the understanding of relationships between variables for the development of hypotheses. Perhaps the oldest and most extensive use of GIS consists of queries for visualizing the distribution of information across space.

Queries tend to be heuristic whereas quantitative analysis relies on rigorous examination of a sufficient amount of sample data. Also called data mining, quantitative analysis reveals patterns in the data that can lead to the development of hypotheses. For example, the USGS uses classification and regression tree (CART) analysis in land cover mapping to explore relationships between spectral response, existing GIS data layers, and land cover classes. Clusters generated from an unsupervised classification are examined with ancillary data including elevation, prior land cover data, census data, city lights data, National Wetlands Inventory data, as well as leaf-off and leaf-on satellite imagery to develop relationships between class occurrence, spectral response, and the other GIS data layers (Hoyt et al. 1998). CART uses a binary partitioning algorithm which recursively splits the data in each node until either the node is homogeneous or the node contains too few observations. The user sets the minimum node size and minimum number of observations used in the split.

Algorithm Specification

Hypothesis development leads to the specification of algorithms that relate inputs to outputs. Model algorithms can be based on expert opinion, quantitative analysis, or a combination of both. An example of expert opinion used in a GIS was recently highlighted in a Discovery Channel program on pacific yew trees. Pacific yew tree bark contains taxol, which has been found to slow the growth of cancerous tumors. Prior to this discovery, Pacific yew trees had little value. Thus, little field data had been collected on the tree. Based on experience, ecologists believed that the trees were located close to streams in old growth closed canopy Douglas-fir stands, and in an elevation band from 2,000 to 4,000 feet. This expert opinion was used to estimate the location of pacific yew trees in the Pacific Northwest. Expert opinion models are based on experience, often are excellent, can be politically powerful, and are also often wrong.

In contrast, models created from quantitative analyses are expensive to build because of the cost of data gathering and analysis. Forest stand growth and yield models and tree volume tables are almost always developed from quantitative analysis. A recent project used GIS to develop a statistical model relating know elk calving locations to habitat characteristics (Bian and West 1997). Over 350 point locations of known calving or non-calving habitat were dropped through several GIS layers to develop the model presented below.

The probability of suitable calving habitat at a particular location = $e^y/(e^y+1)$, where

> Y = -2.583752 + .000706 (distance to gravel roads) + .00034 (distance to highways) - .001867 (distance to seep pits) + 2.222704 (presence of cottonwood) + 2.251458 (presence of cottonwood and salt cedar) + 7.936381 (presence of salt cedar)

Accuracy Assessment

For a model to be useful, its accuracy must be assessed so that the model can be calibrated and so that decision makers can verify the model's reliability. Accuracy assessment is the quantitative measurement and identification of map error. If the results of a GIS model include a map, then its accuracy can be assessed using traditional accuracy assessment techniques (Congalton and Green, 1998). A GIS can be very useful in the selection and allocation of accuracy assessment samples, and in results analysis. In the earlier example of elk calving habitat, a total of 120 accuracy assessment samples were taken across the project landscape. Overall map accuracy was 83 percent.

Future Directions

Since its inception GIS emphasis has centered on creating GIS data layers (e.g. digital elevation models, roads, streams, and land cover). Often 80 percent or more of a project budget is often dedicated to building data layers, with the remaining budget use to develop to one (maybe two) analysis questions. Sensitivity analysis (the process of testing the sensitivity of results to model assumptions) is rare if not non-existent. While GIS allows for spatial analysis, much of the analysis is without the rigor expected of statistics or economic studies. In addition, because of the concentration on data capture, GIS is currently much too centralized. As law firms once had central word processing groups, GIS is too often focused in a small group of experts within a larger organization. Decision makers do not have ready access to the data or to analysis tools specific to their applications.

However, in the last five years, generic data sets have been created for a variety of applications over much of the United States. International layers are also multiplying. As more basic layers become generally available, GIS efforts will shift from data capture towards data access and data analysis. Analysis models will be the focal point of GIS, rather than the afterthought. Future tools will focus on providing solutions to spatial problems rather than providing GIS functionality. Hopefully, spatial analysis and maps will be as common as word processing systems and spreadsheets are now.

References

Abt, R.C. and F.W. Cabbage. *The Sub-Regional Timber Supply (SRTS) Model: Model Structure and Description.* Forthcoming.

Bernath, S., M. Brunego, L. Lackey, and S. Smith. "Using GIS and image processing to prioritize cumulative effects assessment." In *Proceedings: GIS '92 Symposium*, pp. C3, 1-6. Vancouver, British Columbia: Polar's Learning Association, Inc., 1992.

Bian, L. and E. West. "GIS Modeling of Elk Calving Habitat in a Prairie Environment with Statistics." *Photogrammetric Engineering & Remote Sensing* LXIII:2 (1997), 161-67.

Congalton, R. and K. Green. *Assessing the Accuracy of Maps from Remotely Sensed Data.* Lewis Publishers, 1998.

Dana, S.T. *Forest and Range Policy: Its Development in the United States.* McGraw-Hill, 1980.

Green, K. "ISO Ships Its FireLine Data with Pific Meridian Resource's GeoBook." *ArcNews* 19:3 (1998), 26.

Green, K., M. Finney, J. Campbell, and V. Landrum. "*FIRE!* Using GIS to Predict Fire Behavior." *Journal of Forestry* 93:5 (1995), 21-25.

Hoyt, J., et al. *LCCP Land Use/Land Cover Mapping Federal Region 1 Technical Report.* Washington, D.C.: U.S. Geological Survey, 1998.

Landrum, V., B. Carroll, and L. Pious. "Harvest Scheduling in the Age of Ecosystem Management: A GIS Approach." *The Compiler* 14:2 (1996), 6-10.

Vaux, H. "Content of Forest Economics." In W. Duerr and H. Vaux, eds., *Research in the Economics of Forestry,* pp. 17-18. Charles Lathrop Pack Forestry Foundation, 1953.

Spatial Pattern Analysis Techniques

J.M. Klopatek and J.M. Francis, Arizona State University

According to Hobbs (1997) and others, a landscape ecological perspective is necessary for effective natural resource management. Because all ecological processes occur in a spatial context, there is a continual demand for spatial pattern analysis. Once the spatial pattern has been assessed, however, it is necessary to identify the important spatially explicit processes such that predictive models can be developed to assist in resource management. The analysis of spatial patterns in natural systems has progressed dramatically in recent decades due to the rapid development of GIS and remote sensing capabilities (e.g., Coulson et al. 1990).

This chapter does not seek to provide an overview of spatial analysis (Gustafson 1998) nor a listing of quantitative methods (Turner and Gardner 1990). Instead, the intent is to present examples of techniques for analyzing point and patch data, concentrating on woody species of vegetation. These techniques, however, can be extrapolated to different species, habitats, or natural resources at a variety of scales. In brief, these techniques, rooted in the ecological literature, have applications to the analysis of the spatial pattern and management of natural resources.

Landscape Ecology

All resource management fields are related to each other by their dependence on the common science of ecology, a common problem of optimization, and the need for a common set of tools for sampling and statistical and mathematical analysis (Watt 1968). Ecology and plant geography have been largely concerned with the causes of distribution patterns at all scales, from individuals within a small area to vegetation over the surface of the Earth (Greig-Smith 1983). In fact, Andrewartha (1961) stated that ecology was the study of the *distribution* and abundance of organisms. Thus, identifying and quantifying spatial patterns in ecology as a first step in understanding the biological importance of an organism or community has become commonplace. In contrast, the problem of predicting ecological processes at broad scales has remained relatively unresolved (Levin 1993, Gardner and O'Neill 1990).

The investigation of spatial patterns and cause and effect relationships on ecological processes has produced a sub-discipline called *landscape ecology* (e.g., Risser et al. 1984). Landscape ecology focuses on spatial variation in landscapes at a variety of scales, including the biophysical, as well as societal causes and consequences of landscape heterogeneity (i.e., the effect of pattern on processes) (IALE 1998).

As stated above, landscape ecology concentrates on the relationship between spatial pattern and ecological processes across scales (Turner 1989). At the landscape level, spatial patterns in vegetation are generated by coarse scale effects of climate, topography (Swanson et al. 1988), and disturbance (Turner 1989). The resulting pattern of distribution, in turn, affects microclimatic conditions, and mass transfer of energy and resources between the land surface and the atmosphere (Graetz 1991), and the susceptibility of the system to disturbance (Franklin and Foreman 1987, Turner 1987). These coarse scale effects influence finer scale processes such as seed dispersal, germination, competition for heterogeneous resources, differential mortality (Pielou 1961, Brodie et al. 1995), and patterns of resource distribution. Thus, the spatial pattern can be used to examine underlying ecological processes and ongoing biophysical relationships.

Point versus Patch Analysis Techniques

The recent growth in research on spatial dynamics and effects on ecological processes includes a range of ecosystems: temperate and tropical forests (Canham 1988, Lawton 1990, Frelich et al. 1993), grasslands (Wu and Levin 1994), and semiarid shrublands (Phillips and MacMahon 1981, Fidelbus et al. 1996). The spatial pattern of the dominant patch (area) types differs among these systems due to abiotic and biotic factors operating at different scales. At the landscape level the importance of climate in determining community composition has been well documented (e.g., Woodward 1987). At the community level, composition and spatial pattern are further influenced by biological interactions (Greig-Smith 1983, Moeur 1993, Wiens 1985).

There are numerous methods for analyzing patterns in the spatial structure of data (Mueller-Dombois and Ellenberg 1974, Bartlett 1978, Goodall and West 1979, Greig-Smith 1983, Turner et al. 1990). Most methods designed to describe point data, such as stem locations, assume that patterns can be compared to a Poisson distribution, including the variance/mean ratio (Ludwig and Reynolds 1988), nearest neighbor (Getis and Franklin 1989), Ripley's K(d) function (Moeur 1993), and even Morisita's Index (Hurlbert 1990). These methods are primary first-order statistics, which test whether the mean spatial trend is significantly different from a random pattern. First-order statistics distinguish whether the units in a study area exhibit an overall spatial structure (clumped or regular) or are random. However, these statistics are limited in terms of their ability to distinguish between trends and changes in patch configuration.

Second-order statistics (measurement of the square deviation to the mean) were developed for purposes of quantifying small-scale pattern intensity and scale (Fortin, forthcoming). Such methods (e.g., Ripley's I, semivariograms, kriging) test the spatial autocorrelation of variables (everything is related to everything else, but near things are more closely related than distant things) and spatial pattern significance. They are also used to model spatial patterns. (For further discussion of these methods see Rossi et al. 1992, Gustafson 1998, and Fortin forthcoming.)

Techniques designed to analyze point data are not suitable for describing the spatial patterns of habitat patches or canopy cover (Bartlett 1978). The reason is that the Poisson distribution assumes that points, which lack spatial dimension, are spatially independent (Hurlbert 1990). It is theoretically possible, although highly unlikely, that random processes can result in all points being located within an arbitrarily small portion of the map. In essence, patches (landscape elements) may be considered to be aggregations of points and thus, by definition, are not independent because spatial autocorrelation exists. Moreover, the location of a patch on a map will exclude other patches from being located within that region. Consequently, the positions of sites describing habitat patches or canopy cover are not independent and cannot be described by a Poisson distribution. Rasterized data are similarly constrained because only a finite number of pixels can be used to represent a spatial pattern. More than one pixel per point is not feasible.

When discussing spatial analysis of landscape patterns or the quantification of environmental heterogeneity, some definitions are in order.

- *Scale*. The pattern existing in an ecological mosaic–the configuration of different patch types or landscape elements–is a function of scale. Scale represents many phenomena, both conceptual and tangible, including resolution of grain size, sample size and density, extent or area of study, and other relationships in space and time (Forman and Godron 1986, Wiens 1989, Withers and Meentemeyer forthcoming).

- *Fine scale.* Minute resolution or a small study area.

- *Broad scale.* Coarse resolution or a large study area.

- *Grain size.* Minimum sampling unit, or individual units of observations (e.g., pixel size for raster data; Wiens 1989 refers to grain as resolution).

- *Extent. Size* of study area or area included within landscape boundary. In a temporal framework, extent is the time period over which observations were collected. Thus, extent and grain define the upper and lower limits of resolution in a study. Inferences about spatial pattern are constrained by the grain and extent of the study. Therefore, patterns cannot be detected beyond the extent of the landscape or below the resolution of the grain (Wiens 1989).

The spatial pattern of dominant vegetation in a semiarid woodland can be used to illustrate techniques for both point and patch data analysis. Assume that these methods are employed to analyze the patterns of a piñon (*Pinus edulis*)-juniper (*Juniperus osteosperma*) woodland. The pattern of piñon-juniper establishment and recruitment is of interest in documenting the scale of ecological processes such as successional patterns, nutrient cycling, and competition for resources (e.g., water and nutrients). It is noteworthy that information available in point data (i.e., stem locations) does not provide insight into ecological processes such as nutrient cycling, water relationships, and the soil-plant continuum of arid and semiarid landscapes (Klopatek et al. 1998, Schlesinger et al. 1990). Yet, the spatial patterns of the processes are key in modeling vegetation dynamics (Mauchamp et al. 1994).

Point Data Analysis Techniques

Point-mapped data can be analyzed with numerous methods in an equal number of applications. For example, point data have been used to detect departures from randomness (Clark and Evans 1954, Pielou 1959), determine segregation and symmetry between units (Pielou 1961), test relative dispersion (Hamil and Wright 1986), and characterize the spatial distribution of disease (Reich et al. 1991).

Nearest-neighbor Methods

The simplest of point data analytical methods is based on Euclidean distances between individuals, otherwise known as *nearest-neighbor*. In its basic form, nearest-neighbor analysis consists of finding the distance to each unit's nearest neighbor. The distribution of these distances is compared to the expected distribution of distances in a randomly dispersed population. In this form, nearest-neighbor analysis detects a departure from randomness but does not offer any information as to which direction the departure is taking (e.g., regular or clumped), nor does it account for attributes within the population.

Join-count Analysis

Join-count analysis adds another dimension to simple nearest-neighbor analysis in that it allows for the exploration of the relationship between different unit types and can determine whether the dispersion tends to be clumped, regular, or random. This method has been used to analyze spatial patterns of gene frequencies (Sokal and Oden 1978a), tree distribution (Sokal and Oden 1978b), and a plant pathogen system (Real and McElhany 1996).

In a join-count analysis, the location of each unit is recorded as well as an associated attribute such as species, size class, or gender. The nearest-neighbor of each unit is determined by calculating the Euclidean distance between a base unit and all other units. The unit at the smallest distance is designated as the nearest-neighbor (although other criteria, such as distance classes, may be used). The base unit is considered to be joined to its nearest neighbor. It should be noted that this relationship may or may not be reciprocal.

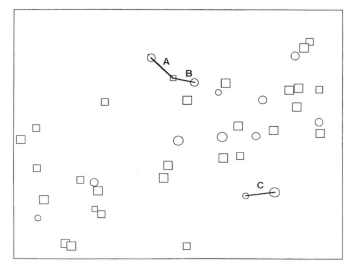

Partial vegetation map of the piñon-juniper woodland showing distribution of piñons (circles) and junipers (squares) and illustrating selected joins (nearest neighbor) relationships.

□ Juniper ○ Piñon

The attributes associated with these neighbors determine the type of join existing between them (e.g., small to large, male to male). The number of each type of join is summed for all pairs of neighbors. Next, the observed number is compared to the number of that type of join in randomly dispersed populations. An excess of any type of join suggests the presence of clumping. For example, an excess of small to large joins suggests that these size classes are aggregated. On the other hand, a lack of that type of join would indicate overdispersion or a negative association between those size classes. Statistical significance of these results can be assessed through calculation of a standard normal deviate (Sokel and Oden 1978a) or, alternatively, a Monte Carlo simulation.

Monte Carlo simulations are accomplished by randomly generating new coordinates for each unit and recalculating the number of joins. Associated attributes, such as species or size, are relocated with the unit. The number of joins is calculated for the randomized data in exactly the same manner as for the observed data. Simulations are repeated until the mean number of joins for all simulations begins to converge on a single value. This value is the expected number of joins given a random distribution. The P value is the number of random simulations that contain fewer joins of that type than observed in the data divided by the total number of simulations. A high P value indicates an overabundance of that type of join in the data, while a low value indicates a shortage. In other words, a high P value indicates aggregation and a low P value indicates overdispersion of the data points. A Bonferroni correction may be applied if testing multiple hypothesis (Neter et al. 1990).

The use of the above method is demonstrated here to characterize the relationship between different woody species along a semiarid environmental gradient in Coconino National Forest, located north of Flagstaff, Arizona (Francis and Klopatek 1998). The study area includes sites representing piñon-juniper woodlands and the transition to ponderosa pine (*Pinus ponderosa*) forest at the higher elevation and Great Basin Desert scrub at lower elevations. The location of each tree and shrub was mapped in numerous 1 ha[-1] sites along with various attributes including species type and size classes. These data were analyzed using join-count statistics coupled with a Monte Carlo simulation to characterize the inter and intraspecific relationships between the different size classes of trees. A Monte Carlo simulation was chosen because it allowed the use of reciprocal pairs of neighbors while ignoring any underlying distribution. Selected results from this study are summarized below.

- The distribution of piñon and juniper trees, with respect to one another, is not significantly different from a random distribution at any site.

- piñon and juniper trees are overdispersed relative to the shrubs in the transition zone between piñon-juniper woodlands and Great Basin Desert scrub.

- piñon and juniper trees are overdispersed relative to ponderosa pine in the transition zone between piñon-juniper woodlands and ponderosa pine forests.

- There is no significant aggregation of size classes in the transition between piñon-juniper woodlands and Great Basin Desert scrub.

- Small trees are aggregated in the two higher elevation sites and this relationship is intraspecific.

- Large trees are overdispersed in the two higher elevation sites but this relationship is not intraspecific.

- Small piñons occur preferentially near large junipers at all sites.

The results of this study indicate that piñon and juniper trees are randomly distributed among each other. That is to say that they do not display either a positive or a negative association at any of the sites. These results suggest that neither inter- nor intraspecific competition dominates the relationship between these species. On the other hand, taken together, piñon-juniper shows a negative association with the other dominant species at the limits of respective ranges. Thus, ponderosa pines are aggregated together (P = 0.991) while segregated from both piñons and junipers (P = 0.044 and P = 0.039, respectively) in the transition between these forests types. This pattern may have arisen from competition for resources (i.e., water or light), although the causes are not clear from this analysis. Likewise, at the lower end of the gradient, shrubs (primarily rubber rabbit brush, *Chrysothamnus nauseosus*) are grouped together between the trees (P = 1.000 for shrub to shrub joins and P = 0.000 for shrub to tree joins). In turn, this pattern is probably a result of competition between the trees and the shrubs for available soil moisture reflecting their similar rooting depths.

The low elevation site contains the lowest density of trees and there is no significant aggregation of size classes present. However, at the other sites, large trees tend to be overdispersed relative to each other (0 < P < 0.023), suggesting the occurrence of either density-dependent mortality or random senescence of individual trees. There is no indication that the species of the nearest neighbor is a factor. Small trees are spatially autocorrelated and the clumps appear to be con-specific (e.g., 0.999 < P < 1.000 for small juniper to small juniper joins and 0.984 < P < .0.999 for small piñon to small piñon joins). In addition, small trees are more frequently associated with large trees than expected from a random distribution. Surprisingly, this relationship is not conspecific. Small piñons occur preferentially near large junipers (0.990 < P < 0.999), while neither small ponderosa pines nor junipers show any preference for large individuals of any species. This result suggests that mature junipers act as nurse plants for young piñon. An alternative possibility is that the small piñon are being suppressed by the junipers.

The advantage of join-count analysis over other types of point based analyses is demonstrated by the previous example. The ability to characterize the relationships between different attributes was very important in this analysis. For example, had the analysis been limited to nearest neighbor, random distribution of piñon and juniper would have been the outcome. Join-count analysis, on the other hand, permitted the characterization of relationships between species and size classes. However, one drawback to join-count analysis is that it is not multi-scale. Therefore, the investigator must use caution in choosing a scale of measurement suitable to the specific research issue.

Lacunarity Analysis

Lacunarity analysis is a spatial analysis method based on fractal theory that provides a scale-dependent description of the spatial patterns of points and/or patches (Plotnick et al. 1993). Mandelbrot (1983) suggested the term *lacunarity* for the distribution of "gaps" or "holes" within a fractal structure. It is important to look at the "holes" of a fractal structure, rather than analyze the fractal pattern directly because elements with similar fractal dimensions may exhibit different spatial patterns.

Allain and Cloitre (1991) describe a simple algorithm for calculating lacunarity that uses a series of "gliding boxes" of increasing size to sample the spatial pattern of gridded data. This method is used to analyze gridded data by placing a "gliding box" with linear dimension of r and area of r^2 at the upper left corner of the grid and counting the number of sites s occupied by the object of interest within that "gliding box." The box is then moved one column to the right and the count repeated. The process continues until all the columns of the top set of r rows have been sampled. The box is then returned to the left edge of the grid, but moved one row down from the starting position.

Note that for all $r > 1$, the boxes overlap and the map is intensively resampled (thus, the sampling units are not independent). Counting continues until all rows and columns have been sampled and the counts summarized as a frequency distribution, $n(s,r)$. $N(r)$, the number of gliding boxes of $2r$ that can be placed on a grid with linear dimension of M, will be equal to $(M - r + 1)^2$, and the values of s will range from 1 to r^2. The frequency distribution, $n(s,r)$, is converted to a probability distribution, $q(s,r)$, by dividing each frequency by $N(r)$. The first and second moments $(Z^{(1)}$ and $Z^{(2)}$, respectively) are calculated as follows:

$$Z^{(1)} = \sum_{s=1}^{r} s q_s \quad \text{and} \quad Z^{(2)} = \sum_{s=1}^{r} s^2 q_s \ .$$

The value of lacunarity, $\Lambda(r)$, for a box with size of $2r$ is estimated by the ratio $Z^{(2)}/(Z^{(1)})^2$. Lacunarity values are similar to the variance-mean ratio (e.g., Ludwig and Reynolds 1988), which can be calculated as $[Z^{(2)} - (Z^{(1)})^2]/Z^{(1)}$, but lacunarity values are dimensionless. The gliding box sampling procedure can be used to estimate lacunarity values for one-dimensional transect data, two-dimensional mapped data, or even three-dimensional data sets (Plotnick et al. 1993). The calculation of the lacunarity index was done using RULE, a spatial analysis model developed by Robert Gardner (forthcoming).

The results of lacunarity analysis are presented here using mapped data sets displaying four different patterns: random, regular, clumped (at 36 units), and a self-similar or fractal map where the pattern repeats itself at all scales within the grain and extent of the map.

Representative maps showing different spatial patterns: A = random, B = regular, C = clumped or aggregated, and D = self-similar or hierarchical.

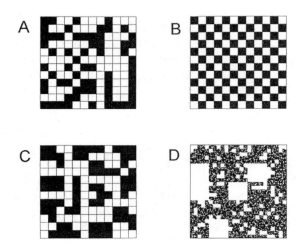

The next figure illustrates the effect of the four different patterns on changes in lacunarity, Λ, with changes in r, the size of the gliding box. Because lacunarity is a function of the occupancy of the sites P (Plotnick et al. 1993), the values of Λ were divided by the value for $r = 1$ to produce "normalized lacunarity" curves, which are plotted on the log scale graph. The simple random map produces a concave curve which shows a sudden drop in the value of Λ as r increases. When r, the size of a side of the gliding box, reaches 10, the value of Λ is approximately equal to 0, indicating that a 10 x 10 quadrat provides a satisfactory estimate of the spatial pattern when occupied sites are randomly located on the map.

The graph of the simple random map also provides a standard against which the other maps can be compared.

Plot of lacunarity values, Λ, using mapped data sets displaying four different patterns as illustrated in the previous figure: random, regular, clumped (at 36 units), and a self-similar or fractal map.

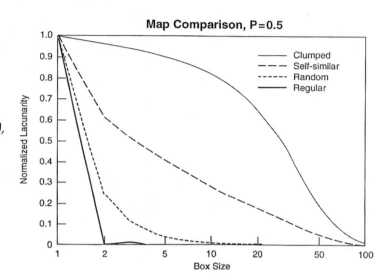

When sites are arranged in a completely regular pattern such as a checkerboard, a sample of size 2 x 2 captures the structure of the entire map ($\Lambda = 0$), and increases in r beyond 2 will add no additional information about the structure of the map. Another way of stating the same is that once the size of the gliding box equals the size of the units from which the pattern is generated then all variability of a perfectly regular map has been captured. A second feature of regular maps is also illustrated—a comparison of lacunarity curves shows that values of Λ for the checkerboard map always lie below those of a simple random map.

The lacunarity curve for the clumped pattern shows a convex curve produced by the random placement of 36 x 36 solid blocks of occupied sites. A comparison of curves shows that lacunarity for the clumped map parallels the slope of the lacunarity curve of the simple random map when r is greater than 30, and approaches an asymptotic value of 0.0 as r increases. The pattern is considered to be essentially random when $r > 36$. The inflection point for this type of curve is determined using the test for concavity from calculus (Swokowski 1979). An inflection point occurs when the sign of the second derivative changes. The second derivative was estimated by calculating the slope for each pair of points on the graph and then calculating the change in slope as a function of box size. Note that when using this method both the slope and the change in slope are approximations and may not be precise. However, because this method worked extremely well with simulated maps, its accuracy and robustness are considered more than satisfactory.

The lacunarity curve for the self-similar map exhibits a continual, nearly linear downward trend to 0.0. A straight line lacunarity plot is thus diagnostic for a fractal or self-similar pattern to the underlying map. That is, the pattern is statistically similar at all scales. If it is assumed that the fractal pattern exists beyond the range of observations, then the structure of the map at either larger or smaller spatial scales can be predicted.

Lacunarity analysis is demonstrated for a piñon-juniper stand, located in the Kaibab National Forest, approximately 25 km southeast of Grand Canyon, Arizona. A one-hectare stand was selected for a burn experiment conducted in 1989 (Klopatek et al. 1992). The site was composed of trees of all age classes, with many trees older than 350 years. The site contained 508 trees of which 340 were piñon and 168 were juniper, with respective cover values of 2,870 and 1,458 m^2 ha^{-1}. All tree stem and canopy locations were mapped out on the ground and an x-y coordinate database was constructed.

The following illustrations show the results of the lacunarity analysis of piñon and juniper stems, with P values of 0.0030 and 0.0014, respectively. Plotted on the graphs is a shaded area representing 10 runs of random distribution at the corresponding P values. Juniper exhibits a completely random stem distribution at all scales while piñon shows a slight trend away from randomness. Because the lacunarity curve is determined by intensive sampling, even small differences can be shown to be statistically significant. The values of lacunarity are great because of large gaps between the stems, but it is not the lacunarity values themselves that are explanatory, but rather the shape of the curves.

Lacunarity analysis of the spatial pattern of piñon (A) and juniper (B) stem locations. The high lacunarity values are reflective of the low cover values of the point locations of the stems, or conversely the large area between the point data. The dashed lines surrounding the solid line represent the output of ten random iterations of maps having the occupancy (P).

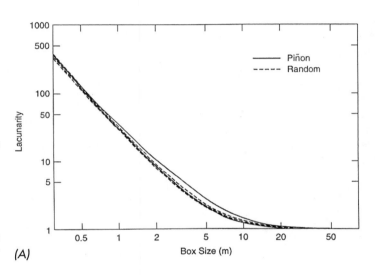

(A)

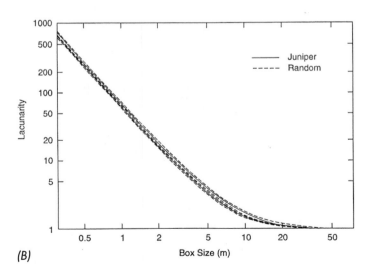

(B)

The next figure presents the analysis of the canopy cover areas. Both species display a clumped, non-random pattern up to a point, with the clumping being greater than that of the average canopy size (12.5m^2), suggesting an association between trees.

Lacunarity analysis of the piñon, juniper, and combined canopy covered patches. The inflection points of the curves indicate the scale of aggregation or clumping (see text).

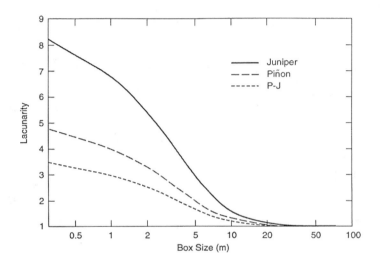

To determine whether the patterns are different, the normalized piñon, juniper, and piñon-juniper combined lacunarity values for cover were overlaid. The results shown in the next figure indicate a remarkable uniformity of pattern for all three canopy cover values. The canopies exhibit a clumping pattern of approximately 35 to 40m^2, distinctly greater then the average canopy size. The clumping pattern corresponds and is supported with the inflection point of the curve at 36.8m^2.

The normalized lacunarity values shown in the lacunarity analysis figure indicate the high degree of similarity in random patterns for the individual and combined tree species.

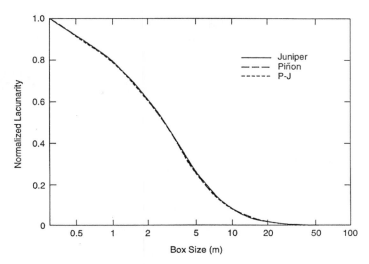

The similarity of spatial pattern displayed by the different tree species in this analysis is remarkable in that stem locations of all tree species at all sites approach a random distribution. Such results suggest that establishment of the trees in this system is a random process occurring against a homogeneous background (Klopatek 1992) and that the random pattern established early in the successional sequence is maintained through later stages.

The lacunarity methodology offers several advantages over other methods of spatial pattern analysis: it can clearly differentiate random populations and deviations from randomness, is valid for point as well as patch (or potentially GIS data), and is insensitive to map resolution and map occupancy (P). The results of a recent study (Klopatek et al. 1998) clearly show the spatial pattern of piñon and juniper stems and canopies, although the empirical distributions were produced by multiplicative processes operating at many scales.

The lacunarity analysis method employed in this study appears to be uniquely suited for analyzing the different or multiscales of patterns, and the method may provide a technique of linking the processes occurring between these scales. A multiscaled analysis is important because it provides a means of defining the scales at which ecosystem processes should be measured and compared. In summary, lacunarity analysis and the gliding box method appear to provide a robust technique for spatial analysis of quantitative data.

References

Allain, C. and M. Cloitre. "Characterizing the lacunarity of random and deterministic fractal sets." *Physical Review* 44 (1991), 3553-58.

Andrewartha, H.G. *Introduction to the study of animal populations.* Chicago: University of Chicago Press, 1961.

Bartlett, M.S. "An introduction to the analysis of spatial patterns." *Supplement Advanced Applied Probability* 10 (1978), 1-13.

Brodie, C., G. Houle, and M. Fortin. 1995. "Development of a *Populus balsamifera* clone in subarctic Québec reconstructed from spatial analyses." *Journal of Ecology* 83 (1995) 309-20.

Canham, C.D. "Growth and canopy architecture of shade-tolerant trees: response to canopy gaps." *Ecology* 69 (1988), 786-95.

Clark, P.J. and F.C. Evans. "Distance to nearest neighbor as a measure of spatial relationships in population." *Ecology* 35 (1954), 445-53.

Coulson, R.N., C.N. Lovelady, R.O. Flamm, S.L. Spradling, and M.C. Saunders. "Intelligent geographic information systems for natural resource management." In M.G. Turner and R.H. Gardner, eds., *Quantitative Methods in Landscape Ecology*, pp. 153-172. Ecological Studies 82. New York: Springer-Verlag, 1990.

Fidelbus, M., R. Francon, and D. Bainbridge. "Spacing patterns in Mojave desert trees and shrubs." In J.R. Barrow, E.D.

McArthur, R.E. Sosebee, and R.J. Tausch, eds., *Proceedings: Shrubland Ecosystem Dynamics in a Changing Environment*, pp. 182-86. U.S. Forest Service, Intermountain Research Station, General Technical Report INT-GTR-338, 1996.

Forman, R.T.T. and M. Godron. *Landscape Ecology.* New York: John Wiley & Sons, 1986.

Fortin, M.J. "Spatial statistics in landscape ecology." In J.M. Klopatek and R.H. Gardner, eds., *Landscape Ecological Analysis: Issues and Applications.* New York: Springer-Verlag, forthcoming.

Francis, J.M. and J.M. Klopatek. "Spatial pattern analysis of shrub and tree species along a semiarid environmental gradient." Unpublished manuscript submitted to *Ecology*, 1998.

Franklin, J.F. and R.T.T. Foreman. "Creating landscape patterns by forest cutting: ecological consequences and principles." *Landscape Ecology* 1 (1987), 5-18.

Frelich, L.E., R. Calcote, M.B. Davis, and J. Pastor. "Patch formation and maintenance in an old-growth hemlock-hardwood forest." *Ecology* 74 (1993), 513-27.

Gardner, R.H. "RULE: A program for the generation of random maps and the analysis of spatial pattern." In J.M. Klopatek and R.H. Gardner, eds., *Landscape Ecological Analysis: Issues and Applications.* New York: Springer-Verlag, forthcoming.

Gardner, R.H. and R.V. O'Neill. "Pattern, process, and predictability: the use of neutral models for landscape analysis." In M.G. Turner and R.H. Gardner, eds., *Quantitative Methods in Landscape Ecology*, pp. 289-308. Ecological Studies 82. New York: Springer-Verlag, 1990.

Getis, A. and J. Franklin. "Second-order neighborhood analysis of mapped point patterns." *Ecology* 68 (1987), 473-77.

Goodall, D.W. and N.E. West. "Comparison of techniques for assessing dispersion patterns." *Vegetation* 40 (1979), 15-27.

Graetz, R.D. "The nature and significance of the feedback of changes in terrestrial vegetation on global atmospheric and climatic change." *Climatic Change* 18 (1991), 147-73.

Greig-Smith, P. *Quantitative Plant Ecology*. Studies in Ecology Vol. 9. Berkeley: University of California Press, 1983.

Gustafson, E.J. "Quantifying landscape spatial pattern: what is the state of the art?" *Ecosystems* 1 (1998), 143-56.

Hamil, D.N. and S.J. Wright. "Testing the dispersion of juveniles relative to adults: a new analytic method." *Ecology* 67 (1986) 952-57.

Hobbs, R.J. "Future landscapes and the future of landscape ecology." *Landscape and Urban Planning* 37 (1997), 1-9.

Hurlbert, S.H. "Spatial Distribution of the Montane Unicorn." *Oikos* 58 (1990), 257-71.

International Association for Landscape Ecology (IALE). "IALE mission statement." *IALE Bulletin* 16:1 (1998), 1.

Kershaw, K.A. "Quantitative and dynamic plant ecology." New York: American Elsevier Publishing Company, 1973.

Klopatek, J.M. and R.H. Gardner, eds. *Landscape Ecological Analysis: Issues and Applications*. New York: Springer-Verlag, forthcoming.

Klopatek, J.M., R.T. Conant, K.L. Murphy, J.M. King, R.C. Malin, and C.C. Klopatek. "Implications of patterns of carbon pools and fluxes across a semiarid environmental gradient." *Landscape and Urban Planning* 39 (1998), 309-17.

Klopatek, J.M., R.H. Gardner, and R.E. Plotnick. "Lacunarity anlysis of the spatial pattern of a semiarid woodland." Unpublished manuscript, 1998.

Klopatek, J.M. "Cryptogamic crusts as potential indicators of disturbance in semi-arid landscapes." In D.H. McKenzie, D.E. Hyatt, and V.J. McDonald, eds., *Ecological Indicators*. New York: Elsevier, 1992.

Klopatek, J.M., C.C. Klopatek, and L.F. DeBano. "Fire effects on woodland flora material and soils in a piñon-juniper ecosystem." In *Fire and the Environment: Ecological and Cultural Perspectives*, pp. 154-59. Proceedings of an international symposium. USDA Forest Service, General Technical Report SE-69, Asheville, North Carolina, 1992.

Krebs, C.J. *Ecology: the Experimental Analysis of Distribution and Abundance*. New York: Harper & Row, 1985.

Lawton, R.O. "Canopy gaps and light penetration into a wind-exposed tropical lower montane rain forest." *Canadian Journal of Forest Research* 20 (1990), 659-67.

Levin, S.A. "Ecological and evolutionary consequences: an overview." In S.A. Levin, T.M. Powell, and J.H. Steele, eds., *Patch Dynamics*, pp.210-212. Lecture Notes in Biomathematics 96. New York: Springer-Verlag, 1993.

Ludwig, J.A. and J.F. Reynolds. 1988. *Statistical Ecology*. New York: John Wiley & Sons, 1988.

Mandelbrot, B.B. *The Fractal Geometry of Nature*. New York: Freeman, 1983.

Mauchamp, A., S. Rambal, and J. Lepart. "Simulating the dynamics of a vegetation mosaic: a spatialized functional model." *Ecological Modelling* 71 (1994), 107-30.

Moeur, M. "Characterizing spatial patterns of trees using stem-mapped data." *Forest Science* 39 (1993), 756-75.

Mueller-Dombois, D. and H. Ellenberg. *Aims and Methods of Vegetation Ecology*. New York: John Wiley & Sons, 1974.

Neter, J., W. Wasserman, and M.H. Kutner, eds. 3rd ed. *Applied Linear Statistical Models: Regression, Analysis of*

Variance and Experimental Designs. Burr Ridge, Illinois: Irwin, 1990.

Phillips, D.L. and J. MacMahon. "Competition and spacing patterns in desert shrubs." *Journal of Ecology* 69 (1981), 97-115.

Pielou, E.C. "The use of point-to-plant distances in the study of pattern in plant populations." *Journal of Ecology* 47 (1959), 607-13.

_____. "Segregation and symmetry in two-species populations as studied by nearest-neighbor relationships." *Journal of Ecology* 49 (1961), 255-69.

Plotnick, R.E., R.H. Gardner, and R.V. O'Neill. "Lacunarity indices as measures of landscape texture." *Landscape Ecology* 8 (1993), 201-11.

Real, L.A. and P. McElhany. "Spatial pattern and process in plant-pathogen interactions." *Ecology* 77 (1996) 1011-25.

Reich, R.M., P.W. Mielke Jr., and F.G. Hawksworth. "Spatial analysis of ponderosa pine trees infected with dwarf mistletoe." *Canadian Journal of Forest Research* 21 (1991), 1808-21.

Risser, P.G., J.R. Karr, and R.T.T. Foreman. 1984. "Landscape Ecology: Directions and Approaches." Illinois Natural History Survey Special Publication Number 2. Champaign-Urbana, Illinois. 67 p.

Rossi, R.E, D.J. Mulla, D.G. Journel, and E.H. Franz. "Geostatistical tools for modeling and intepreting ecological spatial dependence." *Ecological Monographs* 62 (1992), 277-314.

Schlesinger, W.H., J.F. Reynolds, G.L. Cunningham, L.F. Huenneke, W.M. Jarrel, R.A. Virginia, and W.G. Whitford. "Biological feedbacks in global desertification." *Science* 247 (1990), 1043-48.

Sokal, R.R. and N.L. Oden. "Spatial autocorrelation in biology, 1. Methodology." *Biological Journal of the Linnean Society* 10 (1978a), 199-228.

_____. "Spatial autocorrelation in biology, 2. Some biological implications and four applications of evolutionary and ecological interest." *Biological Journal of the Linnean Society* 10 (1978b), 229-49.

Swanson, F.J., T.K. Kratz, N. Caine, and R.G. Woodmansee. "Landform effects on ecosystems and ecosystem processes." *Bioscience* 38 (1988), 92-98.

Swokowski, E.W. *Calculus with Analytic Geometry.* 2nd ed. Boston: Prindle, Weber & Schmidt, 1979.

Turner, M.G. "Spatial simulation of landscape changes in Georgia: a comparision of 3 transition models." *Landscape Ecology* 4 (1987), 21-30.

_____. "Landscape ecology: the effect of pattern on process." *Annual Review of Ecology and Systematics* 20 (1989), 171-97.

Turner, M.G. and R.H. Gardner, eds. *Quantitative Methods in Landscape Ecology.* Ecological Studies 82. New York: Springer-Verlag, 1990.

Turner, S.J., R.V. O'Neill, W. Conley, M.R. Conley, and H.C. Humphries. "Pattern and scale: statistics for landscape ecology." In M.G. Turner and R.H. Gardner, eds., *Quantitative Methods in Landscape Ecology*, pp. 18-49. Ecological Studies 82. New York: Springer-Verlag, 1990.

Watt, K.E.F. *Ecology and Resource Management.* New York: McGraw-Hill, 1968.

Wiens, J.A. "Spatial scaling in ecology." *Functional Ecology* 3 (1989), 385-97.

Wiens, J.A. "Vertebrate responses to environmental patchiness in arid and semiarid ecosystems." In S.T. Pickett, ed.,

The Ecology of Natural Disturbance and Patch Dynamics, pp. 169-93. New York: Academic Press, 1985.

Withers, M.A. and V. Meentemeyer. "Concepts of scale in landscape ecology." In J.M. Klopatek and R.H. Gardner, eds., *Landscape Ecological Analysis: Issues and Applications*. New York: Springer Verlag, forthcoming.

Woodward, F.I. *Climate and Plant Distribution*. Cambridge Studies in Ecology. Cambridge: Cambridge University Press, 1987.

Wu, J. and S.A. Levin. "A spatial patch dynamic modeling approach to pattern and process in an annual grassland." *Ecological Monographs* 64 (1994), 447-64.

Section II

GIS Requirements for Natural Resource Management

Geographic information systems require an appropriate and geospatially accurate framework, various attribute data sets tied to the framework, and metadata to describe how those framework and attribute data sets were created. Because data of every kind are collected to specifications relevant to their intended use, their secondary and tertiary application by others requires an understanding of how errors can creep into the system. An argument articulated by some GIS practitioners states that scale is irrelevant in the digital world because geospatial data can be displayed at any scale desired. There is a counterargument that scale is a fundamental attribute of every data set and that they are best used with similarly scaled data sets. The chapters that follow subscribe to this latter mood, recognizing that such similarly scaled data may not be available for all attributes in every application.

Time, like scale, is another fundamental attribute of data sets. To many practitioners time refers to the age of data. However, for dynamic spatial modeling, frequency of data collection and duration of measurements are the important considerations. Time series analysis allows modelers to play out scenarios to determine whether they achieve *desired future conditions*, or whether there are fractal dimensions in the data set.

Chapter 3 introduces readers to concepts of framework data. The word "framework" is used instead of "base cartographic" because the data sets currently defined as constituting a GIS framework are not entirely the same as those found on cartographic map series developed by the U.S. Geological Survey, National Mapping Division. Resource planners and managers in the United States should be aware that there is a national endeavor to create framework data sets at the finest resolution possible, and that as these become available, they can be electronically generalized to portray larger and larger geographic extents. It is conceivable that someday, resource managers will be able to assemble framework data sets of specific geographies without having to edge match numerous quadrangle maps. The three case studies accompanying this chapter illustrate different dimensions of framework data development, or of the attribute data that constitute framework applications.

Chapter 4 focuses on multi-scale resource data, and particularly on errors that accompany attribute data when they are obtained from multiple sources over long time periods. This chapter also addresses sources of error in field data.

Chapter 5 is a specific look at indicators of environmental and resource conditions. Its aim is to introduce readers to various ways of processing data in a resource GIS to obtain measures not contained in the input data. There are many kinds of indicators present in both spectral and spatial data that can be used to characterize changing landscapes, habitats, and general environmental conditions. The length of road per square mile of land, for example, is a measure of land fragmentation that might well disrupt the habitat of megafaunal species, or signal the rate at which urban fringes are expanding. Similarly, the length of unpaved roads in urban fringes might be an indicator of increased airborne particulate matter that impacts human health.

Multi-scale Framework Data

A. Budge, University of New Mexico

The National Information Infrastructure (NII), or the Information Super Highway, was conceived as a means to make all types of information more accessible on the Internet. Telecommunications, electronic mail, and document retrieval are common uses of the Internet, but spatial data have also become a critical part of the Information Age. They are the essence of modern resource analysis and management. For this reason, the National Spatial Data Infrastructure (NSDI) focuses on distribution and access to geospatial data. To encourage spatial data sharing and to avoid costly duplication of data sets, President Clinton signed Executive Order 12906, Coordinating Geographic Data Acquisition and Access: The National Spatial Data Infrastructure 1995, which directs all Federal agencies to help develop the NSDI and to work cooperatively with other sectors to ensure the infrastructure's growth and evolution. The Federal Geographic Data Committee (FGDC), established by the Office of Management and Budget Circular A-16, is implementing the Executive Order by coordinating the Federal government's sector of NSDI.

FGDC is chaired by the Department of Interior, and consists of 14 Federal agencies and departments. Twelve subcommittees and eight working groups, which bisect all subcommittees, are the foundation of the Committee. These activities are overseen by a Coordination Group, consisting of chairpersons of the subcommittees and working groups. Some of FGDC's higher profile activities include developing and implementing the metadata standards by which users can assess the quality and utility of geospatial data; creating a network of clearinghouse nodes from which users can search and retrieve data; and establishing an infrastructure for framework data development and data sharing. The subject of this chapter is framework data and the infrastructure for developing and distributing these data.

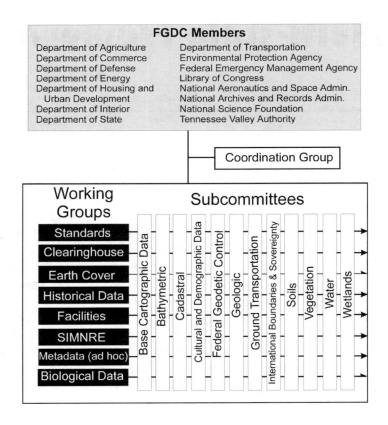

Composition of Federal Geographic Data Committee.

NSDI Strategy

NSDI's vision can be summarized as follows: "[C]urrent and accurate geospatial data will be readily available to contribute locally, nationally, and globally to economic growth, environmental quality and stability, and social progress" (Domaratz 1996). A strategic plan for this vision was developed in 1994 and updated in 1997 to reflect broad community consensus on desirable goals and objectives. The strategy contains the following four main goals, supported by specific objectives.

- Increase awareness and understanding of the vision, concepts, and benefits of the NSDI through outreach and education.

- Develop common solutions for discovery, access, and use of geospatial data in response to the needs of diverse communities.

- Use community based approaches to develop and maintain common collections of geospatial data for sound decision making.

- Build relationships among organizations to support the development of the NSDI.

To realize NSDI goals and objectives, communities and organizations at all levels must work together through cooperative programs, partnerships, and agreements. The NSDI strategy provides a structure under which such relationships can be established. Through the Competitive Cooperative Agreements Program (CCAP) and the Framework Demonstration Project Program (FDPP) FGDC is encouraging NSDI participation and partnering at state and local government levels.

The third goal, in particular, addresses developing organizational relationships and technologies for building distributed, locally maintained data collections. This goal also includes the framework initiative for developing basic data themes onto which other data and applications can be added.

Framework Overview

Goals of Framework Infrastructure

The goals for establishing a framework infrastructure in the United States is to provide the highest resolution base data in a seamless manner to the highest number of users for the widest range of applications; avoid duplication among data collection efforts, particularly at local levels where governmental and management jurisdictions are highly overlapping; and provide a mechanism for high resolution data sets to be updated and maintained to the benefit of all. The framework infrastructure is an electronic counterpart to traditional base cartographic maps that are the current hardcopy equivalents to framework data.

Composition of Framework

Through a series of meetings with representatives from county, regional, state, and federal offices, FGDC developed an institutional and data concept to meet the needs of both data developers and users for cartographic themes most often collected or needed. From these meetings the following framework components emerged: (1) identifying digital geographic data themes and their defining properties; (2) procedures, guidelines, and technology for developing and merging the data sets; and (3) establishing the institutional relationships and business practices required to implement data collection and sharing.

Components and data themes comprising the framework infrastructure.

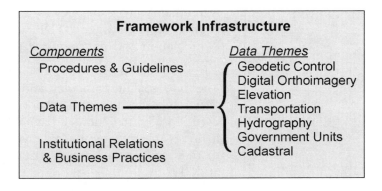

Users nationwide were asked "What data layers do you most commonly use or need as a base for your GIS applications?" The respondents identified the following six common themes as the highest demand data layers in their work: transportation, water, cultural features, elevation, parcels, and boundaries. Not unexpectedly, these mirror the themes most often found on traditional hardcopy base maps. When users were asked if they had a substantial need for better data, the overwhelming response was "Yes!" In this context, "better" usually refers to finer resolution (which often gets confused with "larger scale"), finer levels of categorization, better metadata by which to evaluate the data set, or simply more accurate data.

One recurring need is for common data themes that serve as base layers for other data themes and attributes. Seven data themes have been tagged as framework themes: geodetic control, digital orthoimagery, elevation, transportation, hydrography, governmental units, and cadastral. Referred to as Information Content, these seven theme categories are the focus for building a framework that provides a current base on which to collect and build other data. Used separately or in selectable combinations these base layers (1) define a geographically referenced foundation onto which other details and attribute information can be attached; (2) provide a base map onto which resource data themes can be accurately registered and compiled; and (3) establish a gridded reference map that facilitates compilation of results from analysis of other data.

Framework data must comply with the FGDC Metadata Standard, be dependable, and be constructed from the finest resolution data available, which means they will most likely be developed by nonfederal organizations operating under state and local mandates and responsibilities. Given the diversity of data developers, integration of framework themes across jurisdictional boundaries and between themes is a major issue. These two types of integration are referred to as horizontal and vertical, respectively. Developing the means for achieving these integrations so that seamless data sets for larger areas (that is, progressively smaller scales) can be compiled is the main goal of the framework infrastructure.

Operationally, the infrastructure will address guidelines and technologies for developing, maintaining, and distributing the seven framework themes. Throughout the twentieth century, Earth scientists and natural resource managers have relied on the U.S. Geological Survey topographic map series, particularly the 1:24,000 (or 7-1/2 minute) quadrangle maps for their mapping purposes. As the Information Age evolves, it is apparent that updating these maps on a regular basis is virtually impossible. There are too many dynamic changes occurring at state, county, and municipal levels every day that alter the framework for modern GIS attributes. Thus, it is essential that data collected at these local levels meet minimum standards and collection procedures for internal requirements, but that they can also be integrated horizontally and vertically to meet a larger community of needs. Data themes contributed by organizations at all levels must be verified for accuracy and metadata by area integrators who ensure that data sets can, indeed, be integrated with data from neighboring jurisdictions and with other thematic data sets of the same area. Updates to individual data themes will, by agreement between all contributors, be the responsibility of the original data generating organizations. As the originators and managers of their data sets, these bodies will be asked to perform feature based updates to ensure that other users of their data can be certain they have the best and most recent data for that geographic area.

The framework requires that sound working relationships and business practices be established between participating institutions. These relationships and practices are foundations for building an infrastructure that will create, maintain, and distribute data for a geographic area, and to foster widespread use of these data. Partnerships encourage integrating data produced by local, regional, state, federal, and other sources. New partnerships can be formed to create data required by more than one agency, thereby reducing expensive duplicate efforts. To ensure widespread use of these data, the framework must be responsive to community needs, avoid restrictive policies for data access and dissemination, and provide data at low cost.

Implementation

Implementing the framework infrastructure will be a long-term, evolving commitment for all participants. For the first phase, FGDC is identifying existing local level organizations who already collect high resolution thematic data, and ascertaining their willingness to contribute and maintain these data for the framework. This phase also involves identifying data producers in common geographic areas who, in turn, can stimulate local dialog on the benefits of a fully functional infrastructure, one of which is avoiding duplicate data collection efforts. To be recognized as an FGDC compliant framework theme manager, contributors will be required to provide documentation consistent with FGDC framework theme standards.

Version 1.0, the initial implementation phase, incorporates traditional institutional arrangements and interagency agreements, and establishes new forms of collaboration. The aims of Version 1.0 are to develop and test methodologies for data collection and data sharing arrangements, and investigate requirements to fully implement framework capabilities. Tasks in this phase include researching and developing specifications for the data themes and procedures for data certification and integration. Pilot projects are testing institutional and technical plans. Full implementation is expected to begin before the new century, and will continue evolving data collection and sharing arrangements, and expanding advanced capabilities such as feature based updating.

Data Themes

Each of the seven framework themes is fully documented and contains a limited number of attributes. FGDC's proposed characteristics were published in April 1995 and are being reviewed by practicing GIS specialists through cooperative agreements with the committee. These characteristics may be modified as the pilot projects and framework implementation process progress. The proposed characteristics for each data theme are described below.

Geodetic Control

Geodetic control provides the means for determining locations referenced to horizontal and vertical coordinate systems. It is essential for developing a common coordinate reference for all other geographic features and resource attributes. Control stations are monumented points whose horizontal or vertical location is used as a basis for obtaining locations of other points. These stations and their attributes (name, feature identification code, latitude, longitude, orthometric height, and ellipsoid height) will be included in the framework.

Digital Orthoimagery

An orthoimage is derived from aerial photography or other remotely sensed data that are geographically referenced and corrected to remove displacements caused by sensor orientation and terrain relief. Imagery from the National Aerial Photography Program (NAPP) is commonly used in preparing orthoimages. Orthoimages have a uniform scale and the same metric properties as maps. Digital orthoimagery is an array of geographically referenced pixels that encode values for ground reflectance. Many geographic features can be identified and compiled using orthoimagery. It is also useful as a backdrop and to link attributes of a resource application to the landscape. Imagery that varies from sub-meter to tens of meters in resolution can be included in the framework. Resolutions of one meter or smaller are considered the most useful for compiling framework features.

Digital orthoimage from U.S. Geological Survey digital ortho quarter quad (DOQQ). (Courtesy Earth Data Analysis Center, University of New Mexico.)

Elevation Data

Elevation is a spatially referenced vertical position above or below a datum surface. Elevations of land surfaces and depths below water surfaces (bathymetry) are included in the framework. Requirements for land surface elevations include a regularly spaced grid of locations with elevation values that are collected at post spacings not greater than two arc-seconds, or 155.5 ft at 40° latitude (approximately 47.4m at 40° latitude). A 1/2 arc-second or finer spacing is desired in areas of low relief. Depths below water level will be derived from soundings and a gridded bottom model. Traditionally, a specific vertical reference, usually extrapolated from tidal observations, is used to determine water depths. It is possible that future measurements will be derived from a vertical reference that is based on a global model of the geoid or ellipsoid. This technology is currently used as a reference for determining global positioning system height measurements.

Transportation

For the framework, transportation data include airports, ports, bridges, tunnels, and the centerlines of roads, trails, railroads, and navigable waterways. Attributes for these data are presented in the next table. The purpose in presenting the next table is to reinforce the idea that framework themes should have the fewest attributes possible, and that, because they carry only a feature identification, they can be extracted from the theme, updated as necessary, and reinserted at very low cost.

Attributes for transportation data	
Roads	Feature identification code Functional class Name (including route numbers) Street addresses (encoded as a range between road intersections)
Trails	Feature identification code Name Type
Railroads	Feature identification code Type
Waterways	Feature identification code Name
Airports and ports	Feature identification code Name
Bridges and tunnels	Feature identification code Name

Hydrography

Hydrographic data are based on the approach under development for the U.S. Environmental Protection Agency's River Reach File, Version 3. The Reach File itself, however, is not endorsed as the hydrographic component of the framework. A reach defines a surface water feature that may or may not be connected to other surface water features. Hydrologically connected reaches represent a structure of branching patterns for surface water drainage systems. The framework does not require connectivity and flow direction, although these attributes are desirable. The Reach Code will be the feature identification code. Other attributes are name, reach type, and spatial representation. Reach types are stream, river, lake, pond, wash, or shoreline. Spatial representation refers to spatial elements used to represent the reach (single line, open water, open water shoreline, transport path, junction, or super node). Multiple shorelines are also included.

Governmental Units

Governmental units included in the framework are nation, states (and statistically equivalent areas), counties (and statistically equivalent areas), incorporated places, consolidated cities, functioning and legal minor divisions, federal or state recognized American Indian reservations and trust lands, and Alaska Native Regional Corporations. Attributes for these data are the name and Federal Information Processing Standard (FIPS) code. If the boundary of a governmental unit is also another feature, such as a road or stream, information about this feature will be included, as will the description of the association between the governmental unit and the related feature (e.g., offset, corridor, or coincidence).

Cadastral

Reference systems such as the Public Land Survey System (PLSS) and large publicly administered parcels such as military reservations, national forests, and state parks are considered framework data. Attributes are name and quality; features include the survey corner, survey boundary, and parcel.

Framework Issues

Data developers are challenged with technical and administrative issues whose solutions are essential to make the framework a successful, usable infrastructure. The next table identifies a few of these core issues.

Core technical and administrative issues

Technical	Administrative
1. Coordinate data development	1. Data preservation and archiving
2. Implement standard practices	2. Institutional agreements
3. Define horizontal and vertical integration procedures	3. Data rights and fee structures
4. Implement metadata standards	4. Area integrators
5. Develop feature based updating	5. Data certification
6. Preserve and archive data	6. Data distribution

Technical Issues

Coordinate Data Development

As producers begin to coordinate their data development processes and protocols within the guidelines of the framework, data will be easier to use. However, candidate Framework data developed prior to the infrastructure may not be as easy to integrate. One of the framework goals is to incorporate data from many sources, and at the same time, to recognize the diversity of missions, goals, and resources of these contributors. Thus, one of the primary challenges is to coordinate data development and implement standard practices for developing framework data without imposing on the mandates of the contributors. Each organization has specific applications for the data it develops. For example, data resolution will vary depending on developer requirements. City government employees will develop very high resolution transportation data because they need detail for planning and maintaining street networks. But, a resource manager may require only data that show the general location of major highways. The framework will incorporate multiple resolution data to meet the needs of these differing scales of reference. Data developed at higher resolutions are preferred. Where possible, lower resolution data will be generalized from these high resolution data. To translate this concept to cartographic scale, the framework will include data at 1:100,000 scale or larger.

Implement Standard Practices

The framework requires that a common geographic coordinate system consistent with national and global usage be employed that allows data to be joined and integrated. Latitude and longitude coordinates are recommended because they offer a consistent, seamless system that can be converted to other map projections and grid systems. The North American Datum of 1983 (NAD83) will be used for horizontal reference, and the North American Vertical Datum of 1988 (NAVD88) will reference elevational data.

Horizontal and Vertical Data Integration

Horizontal data integration is a technical issue that requires procedures and protocols to ensure that data will match across jurisdictional boundaries. Because the framework is in its formative stages and horizontal integration rules have not been established, many candidate framework data will not accurately join neighboring data. This does not mean, however, that these data are unreliable. Several guidelines can be followed for using such data, while still recognizing that differences are inherent in the data sets. First, it must be understood that each contributor is the expert for his/her data set, and that other users should not take it upon themselves to change neighboring data to align with their own. Second, logical seamlessness between misaligned data can be achieved by creating a link between the data using a common feature that will be properly coded as such. In the next illustration, for example, road segments A and B do not align properly at the county boundary (dashed line). The preferred approach for addressing this problem is illustrated in inset 1, where a new coded feature (C) is created to link the two segments, rather than assuming segment B is in error and therefore moving it to align with segment A, as shown in inset 2. Albeit a bandaid approach, it allows use of the data until the mismatch can be resolved. Contributors should work together to resolve inconsistencies and to coordinate efforts to create mutually beneficial methodologies for developing truly seamless data.

Linking misaligned segments using a common feature.

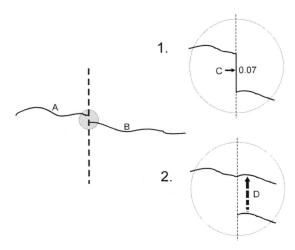

Vertical integration, that is, aligning different data themes on top of one another, contributes to a robust framework. As with horizontal integration issues, existing data most likely will not fit perfectly. The key for successful horizontal and vertical integration is adherence to a common geographic coordinate system.

Implement Metadata Standards

Documenting data is essential. Metadata (data about data) fully describe a data set. Who produced the data? How were they developed? How have the data been updated? How are they distributed? Metadata have the following three main uses: (1) organize and maintain an organization's investment in data, (2) provide information to data catalogs and clearinghouses, and (3) provide information to aid data transfer (FGDC 1995). On June 8, 1994, the FGDC approved the *Content Standards for Digital Geospatial Metadata,* Version 1.0, a required standard for all federally produced geospatial data. Version 2.0 of the Standard underwent public review in 1997 and is expected to be approved by FGDC in summer 1998. Objectives of the metadata standard are to (1) support common uses of metadata, (2) develop metadata from a *need to know* perspective, (3) provide a common set of terminology, definitions, and information, and (4) identify *mandatory* data elements, if they apply, and *optional* data elements.

The standard *does not* specify the means for organizing information in a computer system, the means to organize information in a data transfer, nor the means for transmitting, communicating, or presenting information to users. The standard contains seven main sections with three supporting sections to provide complete information about a data set. The next figure illustrates these sections. Within each of the sections are compound elements and data elements that provide detail for the data set. At first glance, the standard seems daunting. However, when approached section by section, element by element, it becomes more manageable. Software tools have been developed by some public agencies, universities, and vendors to make implementing the standard easier. The FGDC conducted a metadata tool survey in July 1997. A summary of survey results can be found on the FGDC Web site at *http:// fgdc.er.usgs.gov/Metadata/Toollist/metatools797.html.*

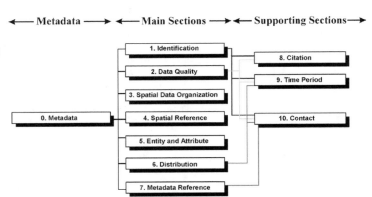

Sections of the content standard for digital geospatial metadata. (Courtesy, FGDC.)

Metadata are also important for providing information about a data set available via the NSDI National Geospatial Data Clearinghouses. Correctly formatted metadata are the key that allow online searching for geospatial data. To encourage use of basic FGDC compliant metadata, a Web based metadata entry and service system is available from the FGDC Web site. This interface is designed for use on UNIX hosts and is meant to collect metadata from users across a local or wide area net to make them immediately available through an NSDI clearinghouse node. Metadata collected in this manner are intended as a first step toward fully implementing the metadata standard and may not provide adequate long-term documentation of data. From the metadata, users can determine if the data set is adequate for their use before acquiring it. Users must remember that metadata alone do not certify that the data are reliable.

Feature Based Updating

Features and feature identification codes are basic to the framework concept. A feature is a description of a geographic object, such as a road. Each segment of the object is assigned a unique, permanent code known as the *feature identification code*. All objects in all themes are required to have these codes because they are used to link framework data with other attributes. A stream, for example, could have different attributes assigned to the same set of feature identification codes, depending on varying applications for the data set. One organization might choose to assign attributes for chemical composition of a stream, another organization might use the data set for attributing numbers of fish, and yet another for relative information on barge traffic. All would use the same framework data set as shown in the next illustration.

Using feature identification codes to attach attributes. (Courtesy, Domaratz 1996.)

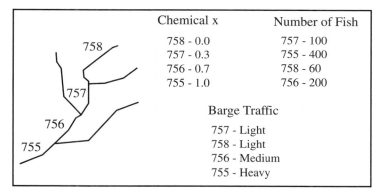

The codes are also the mechanism by which users can perform transactional updates. In the following illustration, line segment A (in the gray circle) can be extracted using the feature identification code and placed into the data set to the right. This is an efficient method for updating data sets shared by many users. Instead of replacing the entire data set, users can update their data by importing the new information using the feature identification code. In addition, the codes provide a link for features represented at different resolutions or scales. The feature identification codes are permanent and will change only if the feature is drastically altered or no longer exists. An example might be a river that is dammed and becomes a reservoir.

Data maintenance using transactional updating. (Modified from Domaratz 1996.)

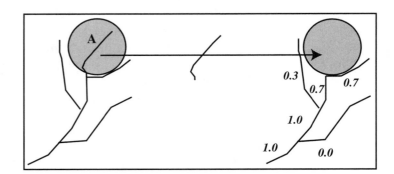

Attributes describe and define a feature's characteristics, such as a name, function, or value. The specific name or value is referred to as the attribute value. For example, in the previous figure showing feature identification codes to attach attributes, the functions assigned by the user for barge traffic on a river are attributes, and each specific function (light, medium, and heavy) is an attribute value. The goal for the framework is to keep attributes for themes to a minimum, thereby allowing flexibility for users to attribute the data according to need.

Archive Data

Historical data are important for temporal studies and will be retained in the framework. Data developers will need to implement a process for preserving these data. The question that most often arises is "How and what do I save?" Archiving digital historical data is an issue that has not been fully addressed. Besides archiving the data, issues of concern include the medium (CD ROM, 8mm tape, etc.), software packages, and file formats to be used. Will current technology be readable in the future? Perhaps changing file formats as new hardware and software are developed is an answer. However, would this practice contaminate the historical value of the data?

Administrative Issues

In addition to technical requirements for developing data, the framework includes an administrative infrastructure that allows the system to function. Administrative issues such as establishing policy, accommodating changing environments, managing the framework, integrating data contributions, producing standards-compliant data, and distributing data must be addressed at management levels. Procedures, protocols, and guidelines for addressing these issues are necessary to make the framework operable. The most desirable model for the administrative infrastructure incorporates participants who have mandates and policies which address these issues. No single organization can, or should, administer the framework. A team of players who have a stake in the framework, its development, and its products should collectively manage the system and be part of the infrastructure. Through a demonstration project in New Mexico, an advisory committee with representatives from all sectors has been formed to begin addressing administrative issues. FGDC has identified six institutional responsibilities (FGDC 1995) discussed below. As the framework is implemented, these responsibilities may be modified.

Policies

To address all administrative components of the framework, the first step is to agree on and establish policies. These policies will be the foundation that set the rules for many of the decisions required of the management team. Practices are required for approving standards, securing funding support, identifying resources, and initiating pilot projects. Because participation is distributed, there will be differing views on how the infrastructure should operate. Partners must agree on how to resolve these differences within guidelines set forth by framework policies.

Theme Expertise

As technology changes and new standards and techniques come into play, data needs of the public and private sectors will change. The framework must be flexible to accommodate these changes. This is an area where FGDC relies heavily upon the expertise of local data developers and contributors that respond to these changes. It is the reponsibility of framework administrators in local organizations and agencies to express needs based on changing trends. They also have the opportunity to drive the responses to these changes. By working together, the GIS community can shape the framework to provide answers that benefit a broad user base at local and regional levels.

Framework Management

One might argue that framework management encompasses all framework tasks and responsibilities. While that is true, FGDC also considers framework management one of the six institutional responsibilities that specifically includes (1) managing production of data themes, (2) managing integration of data themes, (3) recommending and developing technical standards describing data characteristics, (4) developing certification policies and procedures, and (5) ensuring maintenance of a record of data location, and (6) a safe archive. The reader will recognize that these responsibilities overlap others, and that no single organization is expected to assume these responsibilities. These are tasks that should be distributed among organizations and agencies that either already have such responsibilities, or expertise in one or more of these responsibilities.

Area Integration

The framework must be flexible to incorporate newer and better data as they become available. Using the best available data is one of the common denominators. As new data are contributed, they may replace previous data that at the time were the best available. The next figure illustrates such a case. The light gray lines represent roads from the TIGER/ Line files, which at the time were the most reliable data for the area. But, as GPS technology developed, a more accurate road data set was produced for the major roads (the black lines), thereby making it more desirable. Note the differences in accurately representing the road locations, as well as updated information such as the interchange in the lower left.

Area integration also includes accepting new data into the framework and ensuring that they integrate with existing framework data. For example, as counties develop high resolution road data that adhere to the standards and practices of the Framework, these data will be verified by an area integrator to ensure that they are compatible and "line up" with framework road data for surrounding counties. In addition to checking horizontal and vertical integration with other framework data, the area integrator(s) may take on other responsibilities such as implementing technical standards and certification procedures, coordinating data creation and maintenance activities, and providing guidance on integration procedures. Decisions on who will be the area integrator(s) and how the process will take place is left to the communities, states, and regions implementing the framework.

Comparison of road network derived from two sources. (Courtesy, EDAC.)

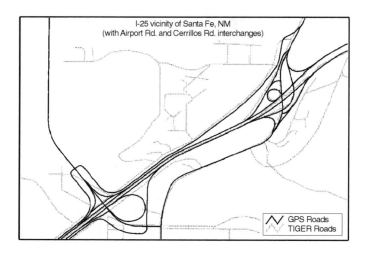

I-25 vicinity of Santa Fe, NM
(with Airport Rd. and Cerrillos Rd. interchanges)

GPS Roads
TIGER Roads

Data Production

Framework data may be developed by a variety of data producers. Local governments (cities and counties) usually produce the finest resolution data because their focus areas are
typically small geographic areas that require very detailed
data. States, regional offices, and Federal agencies manage
larger geographic areas that usually require a more generalized representation. In either case, data producers are
required to develop and maintain data to framework standards, and to provide data and metadata to area integrators.
Adhering to the framework standards includes producing
metadata, performing data quality tests and presenting
results of these tests, and using feature identification coding
schemes in the data set.

Data Distribution

Distribution of framework data is an administrative issue
that can be either a centralized or decentralized activity.
Some communities of developers may decide to make their
data available through a central clearinghouse or service
bureau. This option may be more attractive to local govern-

ment agencies lacking administrative mechanisms for responding to user requests. Other developers may prefer to distribute their own data. In some cases there may be multiple distributors for the same data set. Whatever model is used, the framework guidelines suggest that only one distributor be responsible for holding the "official" distribution copy. In practice this requirement may not be feasible nor desirable. The data developer community must explore options regarding framework data distribution.

Implementing Framework

Realizing the vision of the framework initiative requires cooperation of participants in local, regional, state, and federal government agencies, the private and academic sectors, and nonprofit and nongovernmental organizations. All players have a stake in developing reliable, high quality data on which other data will be draped for resource management decisions for planning, modeling, and analysis.

One of the first steps toward implementing framework is to establish partnerships with agencies having requirements for, or producing framework data. A key factor in building these partnerships is to address the question, *"What is the value of the data to other participating organizations?"* Memoranda of understanding (MOU) or agreement (MOA) may be required to formalize the relationships. Sometimes a common need or a *business case* can be identified that brings these groups together for purposes of creating one of the framework themes. For example, in Michigan three base maps for transportation were developed by different sources over a period of time. Agencies requiring a transportation base selected which map to use, and in some cases used more than one. Multiple databases were developed, resulting in redundant effort. This issue became the business case for the agencies requiring a base transportation data set. A decision was made to utilize a single database of transportation data. As a result of this decision, five agencies joined in partnership to create the data set.

The framework initiative is comprised of many components and is an ambitious program that will evolve over time. However, not all elements must be accomplished at once. Common sense dictates that participants undertake smaller, manageable projects that can grow into larger efforts as the framework progresses. Suggestions on getting started in the process include following best GIS practices such as sharing common data, using common geographic models, using common referencing systems, developing metadata, developing "clean" data, building continuous coverages, and maintaining flexibility. Developing management and business plans, developing partnerships, and identifying supplemental funding resources is advised. FGDC provides an online framework Business Plan Template Kit through the following Web site: *http://www.fgdc.gov*. As with any business venture, a sound business plan is critical. A business plan helps one to (1) develop ideas about how to run a project, (2) document and assess performance, (3) explain the project to others to gain their support, and (4) be dynamic and visionary. The template was developed by people involved in framework projects. The following tables present the business plan template and a template for income and expenses.

Framework business plan template	
Framework Project Name: Geographic Extent: Timeframe: Sponsoring Agencies: Brief Description of Project Overview/Goals: Provide Succinct Narratives for each of the following strategic planning areas.	
Governance	A. What is the organizational structure and governance mechanism for the Framework Project? B. Cite your mission statement, mandate, charter, etc.
Operations	A. Describe your personnel resources, expertise, capacity, and flexibility. B. Describe your technical resources: hardware, software, procedures, tools.
Legal/Political	A. What statutory/regulatory measures impact your project? B. Describe other political parameters or relevant community practices and values.

Framework business plan template	
Products & Services	Describe the framework data and products you offer.
Marketing	A. Who are your customers? What market segments do you address? B. Identify the demand you are addressing: who wants what, and why? C. Describe the price sensitivity and flexibility of this demand, and how your pricing strategy addresses it. D. What is your marketing strategy?
Cooperation	What cooperative goals and opportunities face your project?
Competition	What competitive factors face your project?
Change Factors	What elements in these or other factors will change over the next 3-5 years, and how will they affect your business plan?
Budget	A. Summarize your estimated revenues and expenses for the project's first year in the format provided in the next table. Extrapolate for 3-5 years on separate sheets. B. Describe your fiscal strategies and goals, any quantitative research or modeling you've used, and any strategic assumptions that impact the budget.
Wrap-up	
Attachments	Charts and graphs often help a presentation. Include any that help yours: a chart of your organization, a GANTT or PERT chart that shows project milestones, financial graphs, or others.

Income and expenses template

Income	Data Productn	Data Distributn	Data Mgmt	Coordination	Policy Develop	Resource Mgmt	Monitoring	Total
Appropriations								
Grants								
Service Contracts								
Joint Development Agreements								
Cost Recovery								
Fees and Royalties								
Total								

Income	Data Productn	Data Distributn	Data Mgmt	Coordi-nation	Policy Develop	Resource Mgmt	Monitor-ing	Total
Expense								
Labor Costs								
Contracts								
Supplies								
Equipment								
Indirect Costs								
Total								

Summary

The vision of a nationwide, reliable, high resolution framework of basic geospatial data themes developed by local agencies is an ambitious one. This vision calls for top-down (federally led) and bottom-up (grassroots level) effort, requires long-term commitment by participants that will result in future benefits and cost-saving measures, and requires financial and human resources to sustain the initiative well into the next century. Consider how nice it will be to easily access data that are reliable at very large scales for any location in the United States.

References

Domaratz, Michael. *Common Geospatial Data Framework*. Slide show. Reston, Virginia: Federal Geographic Data Committee, 1996.

Federal Geographic Data Committee. "Development of a National Digital Geospatial Data Framework." Washington, DC: Federal Geographic Data Committee, 1995.

Harmonizing Framework and Resource Data across Political Boundaries: The Tijuana River Watershed GIS

R. Wright, San Diego State University, and A. Winckell,
El Colegio de la Frontera Norte/ORSTOM

Resource Management Requirement

In early 1994 San Diego State University (SDSU) and El Colegio de la Frontera Norte (COLEF) commenced a binational study of the Tijuana River watershed. The project goal was to develop a GIS that could address mutual environmental concerns in a systematic fashion on both sides of the border using data sets that were harmonized across the entire watershed. By late 1994, an agreement was in place for data sharing, coordinating data development and scientific research, and allowing for joint use of the data for education and research (Wright, Ries, and Winckell 1995, Wright and Wright 1995).

The resulting transnational GIS has provided an opportunity to reorient the historical tendency of both countries to develop and manage human and natural resources as though the border were a psychologically and physically impenetrable barrier. Maps printed in each country terminate at the border, and resource policy decisions are implemented without much regard for impacts on adjacent populations on the other side. It is clear that great strides must be made not only in developing new partnerships *within* each country, but also in creating partnerships among resource decision makers *between* the two countries.

The integration process associated with a transnational border surpasses the need to connect disparate databases. Because GIS is a planning tool, its design must integrate diverse values, priorities, cultures, perceptions, electronic conventions (i.e., fostering information access), languages, and legal systems. In the case of the Tijuana River watershed, national, linguistic, and physical barriers magnify these challenges.

This essay provides an overview of the watershed and key framework and resource data developed for or applied to the GIS. Each section focuses on issues and solutions associated with gathering and integrating transnational geospatial data.

The Tijuana River watershed covers 1,750 sq mi, two-thirds of which are in Mexico. It lies astride the California-Baja California section of the U.S.-Mexico border (see the next image). The watershed is a diverse geographical area with a wide range of topography, climates, biological resources, land uses, and sociopolitical institutions. With a combined population exceeding 4 million, the cities of San Diego and Tijuana are particularly important to environmental quality in the watershed and impact virtually every socioeconomic and political question posed regarding the region. The watershed and contiguous areas represent a complex mix of differing agendas, cultures, economic classes, and political systems. Substantial growth is expected within the highly heterogeneous fabric of existing and future development. Much of the growth in the San Diego region is predicted to derive from the North American Free Trade Agreement (NAFTA) and take the form of increased development in the industrial/economic zone along the border. Unfortunately, the growth will occur in an urban system that lacks the basic infrastructure to support such development (Wright, O'Leary, and Stow 1996).

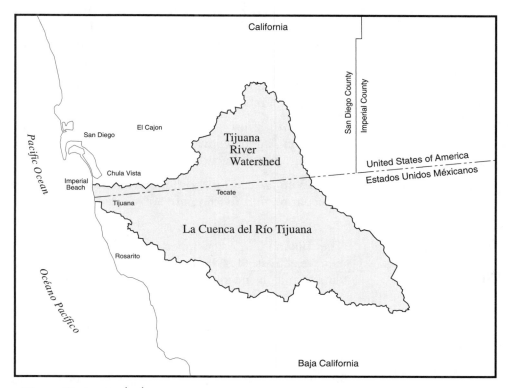

Tijuana River watershed.

Major issues that could be addressed via a harmonized GIS follow.

- *Infrastructure.* Improvements are not keeping pace with population growth and the need for urban services.

- *Natural resources.* The area's natural resources are being depleted through rapid population growth and poor planning.

- *Water pollution.* The semi-arid region is characterized by water pollution and inefficient water usage.

- *Data.* Inadequate data exist concerning the watershed's demographic, economic, and infrastructure characteristics.

- *Communication.* Transborder communications and access to information by educators, residents, researchers, planners, and policy makers are inadequate.

- *Geospatial inconsistencies.* Cross-border geospatial data inconsistencies make it difficult to conduct planning at the watershed level.

- *Coordination.* Little transborder coordination exists among agencies with regional responsibilities for mapping, data collection, and planning.

- *Asymmetry.* Transborder asymmetries in technology, culture, and funding are substantial.

Objectives and Phases

The Tijuana River watershed project has four complementary objectives: (1) GIS database development; (2) use of the database as a means of continuing education and community outreach; (3) providing a means for conducting scientific studies; and (4) enhancement of the capability for integrated watershed management. The project is being carried out in three continuing and overlapping phases. The first phase emphasized the development of framework and resource data. Phase 2 focused on automation of additional data and use of the database for education, public outreach, and scientific research. Phase 3 is focused on improving watershed management.

Development of framework themes and selected resource data were emphasized in Phase 1. A "user needs" workshop, attended by more than 120 individuals, was held in late 1994. The first of several workshops, it was intended to familiarize stakeholders and others with the nature of the project and educate people about GIS. Specifically, the workshop sought to guide the binational team in identifying (1) important planning and environmental issues; (2) types of digital data useful to communities within the watershed and scientific researchers; and (3) the form of products desired by prospective users. The written and oral information obtained from the initial workshop has helped the project team determine priorities for developing and using the database. For example, the team identified numerous research initiatives, some of which are underway in Phase 2. Others (listed below) will begin in conjunction with watershed management activities in Phase 3.

- *Environmental risk assessment.* Rapid urbanization and poor land use practices have created a situation in which the population is at considerable risk from landslides and flooding. Evaluation of environmental risks is underway.

- *Surface water runoff modeling.* The watershed is subject to water flow extremes. An effort is in progress to develop an improved model of surface water runoff to be linked to the GIS.

- *Water and air pollution analysis.* The basin contains numerous generators of water and air pollution. Studies to identify sources of pollution are currently being conducted.

- *Multiple species habitat modeling.* Extensive habitat mapping has been conducted on the U.S. side of the border. Continuing this effort in Mexico recognizes that the border is an artificial biological boundary and that the region should be treated as a single ecosystem.

- *Provision of services.* Urbanization has been so rapid in Mexico's portion of the watershed that the government has been unable to expand services and infrastructure to keep apace with growth. As a first step in evaluating this problem, the team is mapping the distribution of sewage and potable water infrastructures.

Education and public outreach aimed at a wide variety of audiences (e.g., members of the lay public, students of all ages, and professionals from government agencies and the private sector) is an essential component of the project. One of the first products from the database was a two-sided, multicolored poster of the watershed measuring 36" x 46". The poster is composed of several small maps that depict the diversity of the watershed with respect to topography, hydrography, vegetation, climate, and land use. The map on the obverse side of the poster simulates a view from space using a large terrain-shaded rendition of the watershed draped by a false color SPOT satellite image acquired in the summer of 1995. Additional planned initiatives are listed below.

- Video of project and watershed

- One or more interactive multimedia educational modules using ArcView and various parts of the database

- Atlas of watershed and larger area

- Digital flyover through the watershed using aerial photography, satellite imagery, and selected themes from the database.

Finally, the project team aims to incorporate the GIS database into educational and outreach initiatives and scientific studies that in turn lead to a binational approach to addressing environmental degradation in the watershed. A part of Phase 3, this goal will be carried out through such activities as improved land use planning, the consideration of alternate scenarios for sustainable development in the watershed, and effective use of the GIS as a decision support tool for facilitating community involvement in long-range planning.

Harmonizing Framework Data

The key framework data developed for the project are the map quadrangle coordinates, hypsography, hydrography, and roads.

Hypsography and Derived Themes

One of the project's primary accomplishments has been to generate integrated elevation data from U.S. and Mexican sources to create a watershed-wide topographic surface, as well as derived themes such as slope, aspect, shaded relief, and sub-basin boundaries. The derived data were generated in response to specific research needs.

Digital data were derived from scanned U.S. Geological Survey (USGS) 1:24,000-scale contour Mylar separates. Of the 18 quadrangles that comprise the U.S. portion of the watershed, 16 have a 40' contour interval and two have a 20' interval. ARC/INFO software was used to convert the scanned contours to GRID format and then to vectorized contour lines. The contours were edited to correct scanning problems. Next, elevation values and other information such as depression contours, carrying contours, and supplementary contours were added. This process yielded a digitized contour map that meets USGS standards.

The Mexican portion of the elevation database was generated by manually digitizing contour lines from 13 1:50,000-scale maps produced by the Instituto Nacional de Estadística Geografía e Informática (INEGI). These maps have contour intervals of 20m (11 sheets) and 10m (2 sheets).

The final step in developing the hypsography theme was to integrate data from the two sides (see the next figure). Because the DEMs needed to be matched before they could be merged, substantial editing was required. Elevation values on the U.S. side were converted from feet to meters. The combined contour data set was then used as input to the TOPOGRID program in ARC/INFO to generate an integrated 30m DEM. The DEM was then filled to remove processing artifacts and create a topographically correct database.

Hypsography theme.

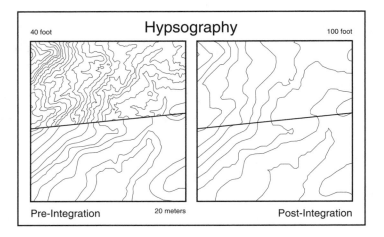

Hydrography

The hydrography layer was generated from the same topographic maps used to generate hypsography. All perennial and intermittent blue line symbols were manually digitized. Given that geology and climate—two factors that greatly influence the characteristics of streams—transcend political boundaries, a significant break in stream density at the border is not expected. However, as seen in the next figure, the INEGI maps contain a much higher density of intermittent stream symbols than USGS counterparts. Clearly, different selection criteria were employed by USGS and INEGI to determine which streams would be portrayed. Minor intermittent streams are not represented on USGS maps, while all intermittent streams are symbolized on INEGI maps. In addition, many blue line symbols did not edge match at the border.

Three steps were developed to create a transborder hydrographic network that would be more or less uniform in density and horizontally integrated across the border. First, the U.S. and Mexican drainage networks were classed according to the Strahler stream order, a system in which each stream segment is assigned a numerical value based on position in the stream network hierarchy (Strahler 1927).

Demonstration of Strahler stream ordering and application to watershed.

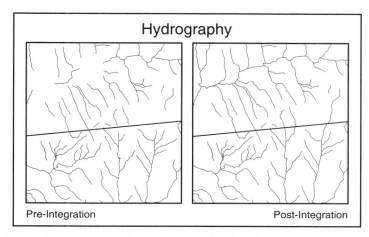

Pre-Integration   Post-Integration

Second, after experimenting with the data, Order 2 and higher numbered segments plus Order 1 streams longer than 925m were selected from the INEGI maps. This procedure considerably reduced the number of short stream segments from the INEGI data set and resulted in a blue line density consistent with USGS maps for the area. Third, blue line symbols were connected at the border where appropriate to correctly represent transborder drainage patterns. The previous image shows drainage patterns before and after transborder line connections were made.

Roads

The road theme was created from existing San Diego Association of Governments (SANDAG) files for the U.S. side of the border and from INEGI maps for the Mexican section. Selecting highways and major streets and roads from the SANDAG file provided an appropriate level of generalization on the U.S. side. For the Mexican side, comparable accuracy and generalization were obtained by updating 1:50,000-scale maps using 1994 1:50,000-scale color aerial photographs and 1995 SPOT 10m panchromatic satellite imagery.

Integration of the U.S. and Mexican coverages involved edge matching the linear features at the three border crossings (i.e., Tijuana-San Ysidro, Mesa de Otay-Otay Mesa, and Tecate-Tecate). To harmonize attribute data, Mexican road classes were reclassified into U.S. classes according to the scheme depicted in the next table.

Equivalent road classes in Mexico and the United States

Class	Mexico	United States
1	Autopista	Interstate
2	Via Pavim 2c	Major highway
3	Via Pavimentada	Arterial
4	Terracería	Unpaved

Data compilation and integration were facilitated through the use of similar projections, datums, and spheroids for original data on each side of the border. Data from Mexico were obtained in ARC/INFO, ASCII ungenerate files for coordinate data, and ASCII files for attribute information. The five-step procedure implemented to create the roads data layer follows.

1. Arc topology was generated from ungenerate files.

2. An empty INFO file was employed to hold the attribute data.

3. The ASCII data for Mexico were placed in the INFO file.

4. ARC/INFO joined the ARC/INFO coverage and INFO file using the Feature ID numbers from the generated topology.

5. A random sample of linear features and attributes was drawn to verify the accuracy of the join process.

Road network theme.

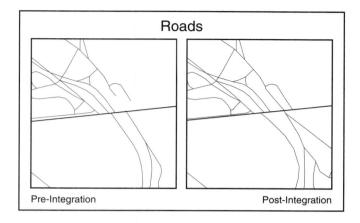

Harmonizing Resource Data

Typically, many solutions can be applied to resolving problems of horizontal geospatial data integration. Simple solutions involve little or no additional data and minimal computer processing time. More complex solutions require more extensive resources. In the case of the Tijuana River watershed, the classification and spatial generalizations employed by Mexico and the United States could be adapted, or a new (i.e., neutral) system not used by either country could be adopted. In either case, issues of time, cost, politics, user requirements, and cartographic generalization must be factored into the final decision. This section illustrates different solutions required to harmonize resource data across an international boundary. Considered here are vegetation, soils, geology, geomorphology, land use, and census subdivisions.

Vegetation

Vegetation for the U.S. portion of the watershed had been mapped and digitized by the city and county of San Diego. This methodology was closely paralleled by SDSU in mapping Mexico's section. Color aerial photographs at 1:12,000-scale were acquired in August 1994. They were used to interpret vegetation boundaries. Minimum mapping units consistent with prior mapping efforts and the ability of mappers to discriminate and delineate features on the photos were chosen. A minimum unit of one acre was selected for wetland and riparian vegetation types because of their relative importance and small size in the region. All other categories were mapped using a 5-acre minimum unit. Individual vegetation types were identified and delineated as polygons on clear acetate overlays placed on aerial photos.

While most mapping was accomplished using photo interpretation in the laboratory, some direct field mapping and a substantial amount of accuracy checking in the field occurred throughout the watershed. The automation process for creating the vegetation layer was performed using on-screen (heads-up) digitizing on a background of rectified, terrain-corrected SPOT imagery. SPOT panchromatic (Pan) and multispectral (XS) images were merged in ERDAS IMAGINE. This technique was effective for integrating and exploiting the high spatial resolution of Pan (10m) data with the multispectral color discrimination advantages of XS (20m) data. On-screen digitizing was performed by SDSU personnel using the ERDAS Vector Module (but may also be achieved using the ARC/INFO Image Integrator). Boundaries were interactively digitized and vegetation categories labeled using the vegetation polygons on the air photo transparencies as guides. Color contrast, edge, and texture enhancement routines in IMAGINE facilitated boundary delineation. Polygons were digitized in the graphic overlay plane of the color monitor using a conventional mouse for location control. An ARC macro language (AML) program was developed to add vegetation category codes and qualifier descriptions. The final step in the automation process was to edge match the Mexican and U.S. portions of the database at the border to create a comprehensive map of the entire watershed (see the next illustration).

Vegetation theme.

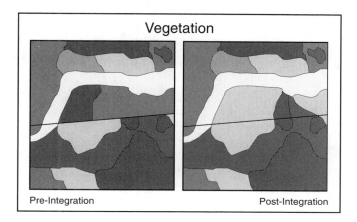

Soils

Soil information is one of the most important thematic layers in an environmental database. An essential component in many models (e.g., those concerned with predicting erosion, surface water runoff, or sensitive habitats), soil data are often unavailable or of variable quality. For the U.S. section, soil data are available on U.S. Department of Agriculture/Soil Conservation Service[1] 1:24,000-scale maps. The data are based on the soil taxonomy classification (U.S. Department of Agriculture 1975) developed by the U.S. government. Numerous levels are recognized in this classification scheme. In descending order (i.e., from most generalized to most specific) the levels are soil order (e.g., alfisols); suborder (e.g., xeralfs); great group (e.g., durixeralfs); subgroup (e.g., abruptic); series (e.g., chesterton); and phase (e.g., fine sandy loam on 2 to 5 percent slopes).

Hierarchical levels of U.S. soil taxonomy

Name	Number of taxa	Differentiating characteristics
Order	11	Degree of "weathering," presence or absence of major diagnostic horizons (e.g., alfisol)
Suborder	52	Wetness, soil moisture regimes, parent material, vegetation effects (e.g., xeralf)
Great group	~230	Characteristics of horizons, soil temperature, moisture regimes, presence or absence of diagnostic layers such as plinthite, fragipan or duripan (e.g., durixeralf)
Subgroup		Intergradations to other great groups, suborders, and orders (e.g., abruptic durixeralf)
Family		Importance to plant growth, textural qualities, mineralogical composition (e.g., fine, loamy mixed thermic)
Series	~12,000	Kind and arrangement of the horizons (e.g., ramona)

On U.S. source maps, soil polygons are differentiated at the series level by a three-letter designator (uppercase, lowercase, uppercase) and sometimes a number (e.g., 2 for eroded and 3 for severely eroded).

On the Mexican side, the only soil data available derive from 1:250,000-scale INEGI maps. Mexico has adopted a classification scheme developed by the United Nations Food and Agriculture Organization (FAO) that recognizes two major classes.

Hierarchical levels of FAO taxonomy

Name	Number of taxa	Differentiating characteristics
Upper class	22 (plus a classification for glaciers and snow caps)	Mainly classified based on conditions under which the soil formed. Some based on diagnostic characteristics such as chemical composition or presence of certain horizons.
Lower class	107 (three upper classes have no lower classes, and one has only a single lower class)	"Composed of intergrades or soils with special horizons or features of note" (Buol 1989:213).

An uppercase letter designates the upper class (or unit) and a lowercase letter the lower class, or subunit (e.g., B = Cambisol, d = Distric). A number is added to indicate soil granularity (1 = coarse, 2 = medium, 3 = fine). In some cases, physical limitations (e.g., rock surface) or chemical limitations (e.g., high salinity) are indicated with additional letters and symbols. On the source maps, taxonomic information is presented as an association between the predominant soil type (shown in first place) and two other associated soil types (shown in second and third places). An example would be: Bd + Hc + Hh/2.

The soil coverage in the GIS effectively illustrates the problems encountered during horizontal integration of categorical data that are systematically different on either side of a jurisdictional border. Strategies are required to harmonize such differences. In the case of the Tijuana River watershed, the first step was to review the two classification systems. The more detailed U.S. scheme was generalized to be more compatible with the FAO scheme. In general, comparability was roughly achieved at the subgroup level of the U.S. classification. However, in some instances comparability was achieved at the series and phase levels.

In the second step, it was necessary to harmonize the reclassified U.S. map with the Mexican map (see the next image). The resulting integration, which is less than ideal, was impacted by three factors. First, the scales and resultant levels of generalization of the soil polygons differ between the two countries. Because of different mapping scales, the U.S. map contains far more detail than the Mexican map. Second, polygons on the U.S. map represent one soil type, whereas polygons on the Mexican maps may represent one, two, or three associated soils. Third, soils in urban areas are not portrayed on the Mexican map but they are on the U.S. map, albeit as an urbanized phase.

Soils theme.

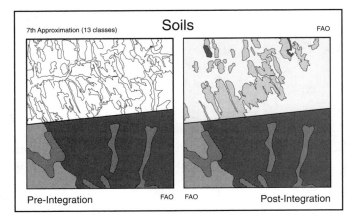

One of the watershed project's future tasks is to obtain a more detailed delineation of soils on the Mexican side by correlating them with other themes in the database such as slope, elevation, climate, vegetation, geology, and geomorphology. However, ultimately the production of a uniform soil map of the region will require agreement between U.S. and Mexican soil scientists on a common soil classification for the border region and the subsequent mapping of soils using similar techniques, decision rules, and levels of generalization.

Geology

The geology layer was derived from existing U.S. and INEGI paper maps. For the California section, the reference map was the Geologic Map of California, San Diego-El Centro Sheet at 1:250,000 scale (State of California Resources Agency, Department of Conservation, Division of Geology, Sacramento) updated from 1962 to 1992. The Mexican source map, also scaled at 1:250,000, was entitled "Geologic Map, Tijuana, 111-11" from INEGI (1982-83). One classification was developed from these maps using age, origin, surrounding formations, lithology, and associated facies as criteria. The integration into a single spatial layer involved resolving discrepancies in nomenclature and different cartographic approaches. Additional photo interpretation and field verification, along with other information in the GIS (namely, the geomorphology theme), helped solve these issues.

Geomorphology

The geomorphology theme for the project involved original mapping in the field by researchers at COLEF and the Centro de Investigación Científica y de Educación Superior de Ensenada (CICESE). They followed a geomorphological classification developed by Zink (1988-89). The classification is multicategorical and hierarchical, with six levels of categories and increasing generalization at higher levels. For this database, the landscape and landform levels were used.

Landscape is defined as a place on Earth characterized by repetition of different types of landforms (see the following table). The morphological classification used at the landscape scale is based on elevation and slope of the terrain computed from the project DEM.

Terrain classification at the landscape level

Form	Slope (%)	Class
Flat	0.0-1.9	Plain, valley
Gently undulating	2.0-7.9	Peneplain, piedmont
Undulating	8.0-15.9	Low hills
Dissected	16.0-29.9	High hills
Highly dissected	30 and above	Mountains

Landform is defined as a surficial form of the Earth identified from data on topography, geological structure, morphogenetic processes (e.g., cuesta, glacial fan, terrace, or delta; see the next table). At this level, genetic and structural characteristics are more significant. The same morphological agent could result in different forms depending on the context in which the action takes place. For this reason, a distinction is made between erosional and depositional forms. For example, the unit "depression" is listed under both erosional and depositional processes.

Structural	Erosional	Depositional	Residual
Mesa	Depression	Depression	Erosional surface
Dome	Canyon	Fluvial plain	Inselberg
	Hill	Glacial deposit	Tors
		Gentle slope	Terrace
		Steep slope	Alluvial fan
		Summit	
		Undulating slope	

The landform/landscape units in these tables were integrated from 1:50,000-scale color aerial photographs and the project DEM, and then integrated at the border.

Geomorphology theme.

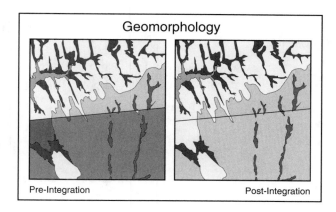

Land Use

Twelve classes of general land use and land cover for the Mexican portion of the watershed were interpreted from 1:12,000-scale National Oceanic and Atmospheric Administration (NOAA) photographs and checked in the field. These classes are the same as those developed and employed by SANDAG for its land use map of San Diego County. The land use polygons for the Mexican side were placed in digital form using heads-up digitizing against the background of a merged SPOT panchromatic and multi-spectral image that had been terrain corrected and georeferenced. These digital data were then merged with general land use for the California section of the watershed previously created by SANDAG. The result is a seamless land use coverage of the entire watershed.

Land use theme.

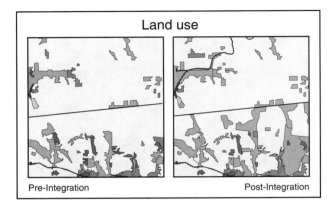

Land use

Pre-Integration   Post-Integration

Census Enumeration Units

The driving force behind much of the environmental change in the Tijuana River Watershed is the attempt to provide both housing and services for a rapidly increasing population. To establish measures of housing and demographic characteristics, the two national governments have divided the countries into enumeration units for data collection purposes. These units, called *áreas geoestadísticas básicas* (AGEBs) in Mexico and census tracts in the United States, vary greatly in geographic size, shape, and the type of enumerated data collected. To unite the files, ungenerate graphical files were merged into a single ARC/INFO coverage. Edge matching was unnecessary along the shared boundary; however, the international border was inspected to ensure that no overlap of polygon features occurred.

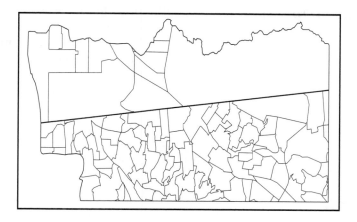

Census tract/AGEB theme.

Summary

The binational approach used to develop the Tijuana River watershed GIS reveals previously overlooked common issues as well as the barriers that have prevented the establishment of a single approach to spatial problem solving. The binational public mandate to focus the GIS on matters pertaining to water quantity, quality, treatment, and use has already begun to forge new partnerships, identify conflicting perceptions, spotlight issues that transcend the international border, develop a constituency in support of cross-border environmental issues (such as coastal sage scrub and estuarine areas), and identify and access data that have largely been unavailable to researchers and the public. Binational work on the GIS has assisted the growth of a grassroots community required to develop cross-sector and binational trust. This watershed GIS is starting to involve the public in articulating common issues and management priorities.

Integration of geospatial data across international borders involves many technical and human issues. In general, solutions to technical problems are easier than those relating to human and cultural factors such as legal and organizational impediments to data.

Acknowledgments

The authors gratefully acknowledge the start-up funding for the Tijuana River Watershed Project provided by the National Oceanic and Atmospheric Administration as well as subsequent support from the United States Department of Defense, the United States Geological Survey, the United States Environmental Protection Agency, the Southwest Center for Environmental Research and Policy, and the San Diego Association of Governments. We also wish to recognize the contributions of Laura Edwards and Harry Johnson for their efforts in creating the maps for this article.

Endnote

1. The SCS has been renamed the Natural Resource Conservation Agency (NRCA). Among other functions, it is responsible for mapping and describing soils of the United States, and for making these data accessible to the public in GIS compatible electronic form.

References

Buol, S.W. *Soil Genesis and Classification*, 3rd ed. Ames, Iowa: Iowa State University Press, 1989.

Strahler, A. "Quantitative Analysis of Watershed Geomorphology." *Transactions of the American Geophysical Union* 8:6 (1927), 913-20.

U.S. Department of Agriculture, Soil Survey Staff. *Soil Taxonomy: A Basic System of Soil Classification for Making and Interpreting Soil Surveys*. U.S. Department of Agriculture Handbook, No. 436. Washington, D.C.: U.S. Government Printing Office, 1975.

Wright, R., K. Ries, and A. Winckell. *Identifying Priorities for a GIS for the Tijuana River Watershed: Applications for Land Use Planning and Education*. Workshop Proceedings, Institute for Regional Studies of the Californias. San Diego, California: San Diego State University, 1995.

Wright, R. and P. Wright. "The Tijuana River Watershed GIS Project: A Binational Consortium." Proceedings, GIS/LIS 1995, Nashville, Tennessee, November 1995, pp. 1046-55.

Wright, R., J. O'Leary, and D. Stow. "The U.S.-Mexico Border GIS: The Tijuana River Watershed Project." In Stan Morain and Shirley López-Baros, *Raster Imagery in Geographic Information Systems,* pp. 322-28. Santa Fe, NM: OnWord Press, 1996.

Zink, J.A. "Physiography and Soils." Lecture notes. International Institute for Aerospace Survey and Earth Sciences, Enschede, The Netherlands, 1988-89.

A Modular Ground Water Modeling System

J. Yan, South Florida Water Management District; K.R. Smith,
Thurston County Public Health Department; and R.M. Greenwald,
P. Srinivasan, and D.S. Ward, HSI GeoTrans

Resource Management Requirement

To support water resources evaluation, planning, and regulation, the South Florida Water Management District develops and maintains a ground water modeling information base consisting of calibrated regional ground water models and associated data sets. Frequently, new models—typically at larger scales and with finer grid resolution—must be created to analyze local impacts of wellfields, water control structures, and water management alternatives for water supply, flood control, and environmental enhancement. To facilitate the creation of these larger scale models, a modular ground water modeling system called "GWZOOM" was created using a GIS.

GWZOOM provides an interactive environment for generating georeferenced model grids on a computer screen. Spatial and temporal data can be distributed among all grid cells and automatically assigned to model coordinates by layer, row, and column. The model input is created in the format required by the ground water flow model code (MODFLOW). GZWOOM enables modelers to create, apply, and revise ground water models quickly, and it constitutes a new system for organizing, processing, and storing data for ground water models.

GWZOOM was used to create a large-scale model from a regional scale model in Dade County, Florida, in order to analyze a proposed underground seepage barrier known as a "curtain wall." The purpose of the wall is to increase water levels and improve hydroperiods in the Everglades National Park while protecting adjacent agriculture from floods and other harmful events. GWZOOM was used to generate the model grid, convert regional model data sets to the larger scale model, and transfer data from various coverages in the GIS to MODFLOW coordinates and input files. Many alternative scenarios were also created and simulated. Models were generated and revised far more rapidly than previous processes allowed.

GWZOOM Concept

The GWZOOM concept is based on the modular three-dimensional finite-difference ground water flow model known as MODFLOW (McDonald and Harbaugh 1988). Typically, several model layers are employed to represent aquifers. The geologic formations are partitioned areally and vertically to include spatial variability, hydraulic properties, and boundary conditions. All model input and output are associated with the model grid coordinates (layer, row, and column) at the node (center) of each grid cell. GWZOOM has the ability to automatically process data from GIS coverages and data sets from regional ground water models into a set of new model coordinates representing a larger scale grid.

Aquifer partitioned in finite-difference grid.

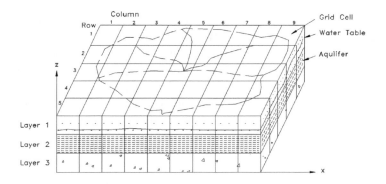

To accomplish this, ARC/INFO software is used to (1) partition the hydrogeologic system, (2) invoke the overlay commands to intersect coverages with a grid coverage, and (3) automate the assignment of hydraulic parameters and boundary conditions to the zoom model grid coordinates (layer, row, and column). A grid polygon coverage must be created to represent the finite-difference grid of the zoom model within a regional model area. Once the data for the regional model are in a GIS format, the overlay commands produce new coverages with a parameter value for each grid cell in the zoom model. The report generating commands (using ARC/INFO's AML, or Arc macro language) can be used to generate ASCII files, and FORTRAN or C subroutines are used to convert the data in the files to the format required by MODFLOW.

GWZOOM System

The system is operated from within ARC/INFO and can be started by typing *gwzoom* at the ARC prompt. The GWZOOM main menu appears on the screen, providing the options below.

Command	Action
GWArc	Performs model related GIS operations in an ARC/INFO workspace.
GWAlt	Provides major alternative options for creating model inputs, analyzing model results, and adding new modules.
GWPre	Performs model related file compilation and form driven input for model parameters.
Execute	Runs a MODFLOW model for which all input files have been prepared.
Exit	Exits GWZOOM and returns to ARC/INFO.

GWArc and GWPre were programmed with modular structures and share common subprograms corresponding to all modules of MODFLOW, while GWAlt was programmed with independent modules for special needs and alternative operations (ESRI 1990).

Implementation

GWZOOM was used to model the seepage barrier proposal (see the next illustration). The Dade County regional model is one of a series developed by the South Florida Water Management District to support planning and regulatory activities. Its grid consists of 132 rows and 100 columns with a uniform cell size of 0.5 mi x 0.5 mi. In descending order, the grid has four layers representing the Biscayne aquifer (Miami Oolite Formation and Fort Thompson Formation) and Tamiami Formation (low permeability layer and gray limestone aquifer). The larger-scale zoom model (called the "Frog Pond model") has 150 rows and 200 columns with a uniform cell size of 500 ft x 500 ft. Layers in both models are the same.

Frog Pond model area.

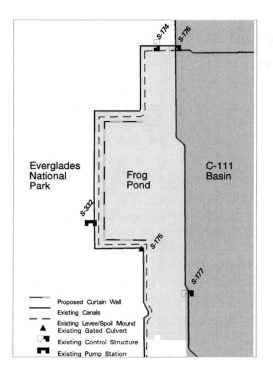

Boundary conditions for the Frog Pond model were defined as general head boundaries, with the head values interpolated from the regional model for each one-month stress period. This approach provides the Frog Pond model with fine-scale simulation capability for the study area while maintaining consistency with the hydrologic conditions of the regional flow system simulated by the Dade model. The computer code used for both models was MODFLOW.

Major GIS coverages for the Dade County model are listed in the table below. The canal stage data were assembled into an ASCII file with control structure names identical to those in the canal GIS coverage for linking the static data with temporal data. The Dade model was run to generate hydraulic head values for establishing the local model's boundary conditions.

GIS coverages for the Dade County spatial database

Coverage name	Brief description	Coverage type
Land-use	Land use map in level 3 detail	Polygon
General-soils	Soil types and hydraulic parameters	Polygon
Land-surface	Topographic land surface	Line or point
Base-map	Major roads, canals, political, and hydrologic boundaries	Polygon and line
Canals	Canal location, width, bottom elevation, associated control structures, river or drain classification, and other attributes	Line
Layer-Bottom	Bottom elevations of model layers	TIN*
Conductivities	Hydraulic conductivity parameters	TIN
Transmissivity	Transmissivity parameters	TIN
Storage-Coeff	Storage coefficients	TIN
Vcont	Vertical conductance	TIN
Rain-station	Location, ID, and static data	Point
Grid	Grid coverage coincident with model grid	Polygon
*Triangulated irregular network.		

The zoom grid was created using the MODEL GRID option of GWZOOM. Once the grid is created, the MODEL DATA option is used to create data files for each MODFLOW package. The following example is provided to describe the operations.

1. To create the data file for hydraulic conductivity of layer 1, the zoom grid and TIN of hydraulic conductivity of the regional model are used. On the MODEL DATA menu (refer to the following figure), the user inputs the name of the GIS workspace containing the zoom grid, the name of the modeling workspace to which the MODFLOW files are written, and the model run.

2. The user selects the BCF package (a matrix-based package), parameter "hyc" (for hydraulic conductivity), model layer 1, and clicks on the create/edit button.

3. A new menu appears, providing the user with four button options for creating a GWZOOM coverage for matrix data, one of which is to spot an existing TIN. The user clicks on this button, and then specifies the name of the TIN. GWZOOM creates a GWZOOM coverage called *g-hyc-1-1a* according to a systematic naming convention developed for GWZOOM. The GWZOOM coverage for matrix data is always a point coverage with one point at each model grid center and attributes identifying the model rows and columns.

4. After the coverage is created, the screen automatically returns to the MODEL DATA menu. The user then clicks on the write button and GWZOOM writes an ASCII file named *hyd-1-1a.mod* from the coverage *g-hyd-1-1a*. This file contains the hydraulic conductivities of layer 1, for run 1a, in a matrix format readable by MODFLOW. Following the same procedure, all data files can be quickly created for a zoom model.

MODEL DATA menu.

```
┌─────────────────────────────────────────────────────────────────────┐
│  ⌐                           MODEL DATA                                │
│        GIS workspace:   ( change )   /net/somerset/usr3/jyan/colruns/crewzoom/gisdata
│   Modeling Workspace:   ( change )   /net/somerset/usr3/jyan/colruns/crewzoom/modeldata
│           Model Run:    ( change )   Run 2a
│  ═══════════════════════════════════════════════════════════════════
│                             PACKAGE:
│        LIST—ORIENTED                          MATRIX—ORIENTED
│        ─────────────                          ───────────────
│                                            BY LAYER:
│        _|    River
│        _|    Drain                         _|      Basic
│        _|    Ghb                           ✔       Bcf
│        _|    Well                          _|      Generic By Layer
│        _|    Pws
│        _|    Aws
│        _|    Obs                           BY STRESS PERIOD:
│        _|    Canal
│        _|    Lake                          _|      Recharge
│        _|    Generic List                  _|      Et
│                                            _|      Generic By Period
│  ═══════════════════════════════════════════════════════════════════
│   PARAMETER:                               FOR MATRIX DATA ONLY:
│     Code:   hyc                              Layer:    1
│     (3—letter code)                          (use "0" for all)
│  ═══════════════════════════════════════════════════════════════════
│   DATA TYPE:                               OPERATIONS:
│        _|    Integer Matrix                ( create/edit )  ( arc prompt )
│        ✔     Real Matrix                   ( MTV link )     ( help )
│        _|    List                          ( write )        ( ok )
└─────────────────────────────────────────────────────────────────────┘
```

Results

Both steady state and transient simulations were created for the Frog Pond model. The transient simulation covers January 1988 through December 1988, which includes both average and wet conditions.

The historic ground water table in the area has a northwest to southeast gradient, with water table elevations similar to canal stages. The historical (1933-1947) canal stages vary from 1.9 ft to 5.6 ft. Since 1947, canal stages have been maintained between 2.7 ft and 4.0 ft to facilitate flood prevention and a stable water supply. For purposes of environmental restoration, it has been determined that canal stages and ground water levels should return to historical levels, or 1.9 ft to 5.6 ft. However, the agricultural interests in the area require a relatively stable ground water level during the growing season. Extremely high or low water levels have a negative impact on agricultural areas to the east. The curtain wall concept was proposed to solve these problems by hydraulically separating the two areas.

The first image in the next set of figures shows the simulated ground water level contours with the proposed curtain wall. The bottom image shows the water level contours simulated without the proposed curtain wall. The two figures indicate that the water levels in the Frog Pond area are reduced slightly by the wall. Apparently, only the 3.44-ft contour line is moved north about 1 mi. This simulation reveals that the curtain wall is not very effective under these assumed conditions.

Water level contours simulated with a curtain wall.

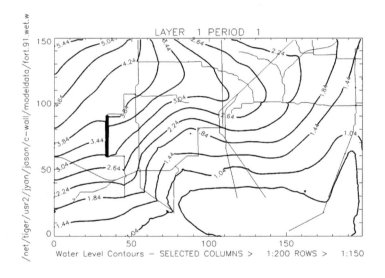

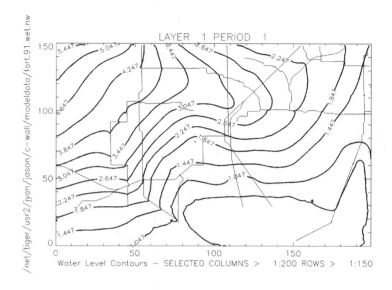

Water level contours simulated without a curtain wall.

Conclusions

The GWZOOM system integrates four major components for developing and applying models: (1) a georeferenced hydrologic database, (2) a data processing and transformation system for creating scalable ground water models, (3) a ground water simulation program, and (4) a system for graphically displaying and analyzing model input and output.

Implementing the GWZOOM modeling system provides the following advantages.

- The graphical interface allows the user to easily create and apply a local scale model based on a regional model.

- The system makes it simple to understand and prepare model components.

- The system's generic structure can be applied to different study areas or MODFLOW models.

- The system's modular structure is easily enhanced with additional modules.

Although full implementation of the system requires converting data sets for existing regional models into GIS compatible formats, this can also be quickly and easily accomplished using options included in GWZOOM.

References

Environmental Systems Research Institute. *Understanding GIS: The ARC/INFO Method, Version 7 for UNIX and Open-VMS*. Redlands, California: ESRI, 1990.

McDonald, M.G. and A.W. Harbaugh. "A Modular Three-dimensional Finite-difference Ground Water Flow Model." Techniques of Water Resources Investigations of the United States Geological Survey, Book 6. Washington, D.C., 1988.

Locating the Continental Divide Trail in New Mexico

K.A. Menke and K. Mich, University of New Mexico

Resource Management Requirement

The Continental Divide National Scenic Trail (CDNST), which traces the divide from Mexico to Canada, became part of the National Trails System in 1978. When complete, it will span 3,100 mi and pass through five U.S. states, 25 national forests, 16 wilderness areas, three national parks, one national monument, four Bureau of Land Management districts, and thousands of acres of private land, as well as several historic sites (see the next illustration). By 1978, much of the route from Colorado north to the Canadian border existed in the form of individual trails and dirt roads.

Challenges in New Mexico

In New Mexico, however, most of the trail had yet to be constructed or placed because the divide passes through Native American reservations, Hispanic land grants, and other private property. The U.S. Forest Service (USFS), Bureau of Land Management (BLM), National Park Service (NPS), and two grassroots trail organizations have been working together to route the trail through the state. Progress has been slow as a result of insufficient communication among the federal agencies and grassroots entities.

Students at the University of New Mexico's Department of Geography have proposed using GIS and GPS technologies to identify the best possible trail route. Representatives of the federal agencies and private trail organizations could then view alternatives in the context of private land ownership, land use, topography, existing roads, and available water sources. With these data as a framework, trail managers would be able to create the best possible trail for all concerned.

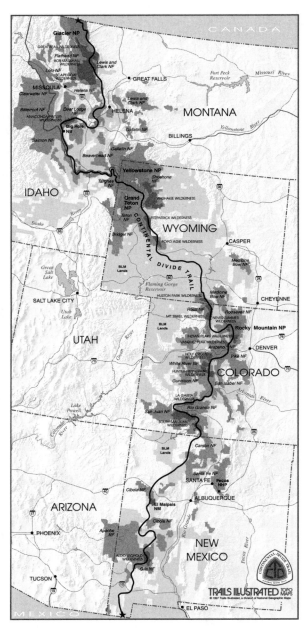

Current Continental Divide Trail. Courtesy of Trails Unlimited Topo Maps.

Enabling Legislation

On Oct. 2, 1968, Congress passed the National Trails System Act. This act "established policies and procedures for a nationwide system of trails consisting of National Recreation Trails, National Scenic Trails, National Historic Trails and connecting or side trails." The act designated the Appalachian Trail and Pacific Crest Trail as the first such trails, and authorized study of a Continental Divide trail. The act specifies that national trails be established primarily for hiking and horseback use. Motorized vehicular traffic is specifically prohibited. This restriction could have had grave consequences for trail location in New Mexico. However, in 1983 an amendment was passed that allowed use of motorized vehicles if such usage existed when the trail was designated. This amendment also permitted trail activities such as bicycling, cross-country skiing, jogging, snowmobiling, and recreational water activities.

After a decade of study, the Continental Divide Trail was officially designated. On November 10, 1978, Public Law (PL) 90-543 was amended as PL 95-625, which established the trail as part of the National Scenic Trails System. According to its mandate, the Continental Divide Trail is to be an "extended trail located so as to provide for maximum outdoor recreation potential and the conservation and enjoyment of the nationally significant scenic, historic, or cultural qualities of the areas through which it passes."

Multiple Interests

Government Agencies

Most of the trails in the National Trails System, including the Appalachian and Pacific Crest trails, are administered by the National Park Service (NPS). However, the lead agency in managing the Continental Divide Trail is the U.S. Forestry Service (USFS) because it manages the largest share of land along the corridor. The BLM and NPS are responsible for designating, building, and maintaining sections of trail passing through respective holdings. In addition to the BLM, USFS, and NPS, there are two grassroots organizations working on developing routes for the trail: the Continental Divide Trail Society and the Continental Divide Trail Alliance.

Continental Divide Trail Society

The Society was establish in 1978 by Jim Wolf. Known by many as the father of the Continental Divide Trail because of his active lobbying, Wolf spoke at the oversight hearings before the Subcommittee on National Parks and Recreation in favor of a Continental Divide trail in 1976. He formed the nonprofit organization two years later in order to step up lobbying pressure. A Society brochure describes the trail as a "silent trail, one laid out with appreciation of its natural environment and sensitivity to yearnings for a sense of contact with the wilderness." Applied to New Mexico, this description yields several viable alternatives for sections of the trail that federal agencies have decided to place on roads or route away from areas of scenic interest.

Continental Divide Trail Alliance

The Alliance was formed as a nonprofit organization in October 1995 by the Fausel and National Forest foundations. Funded by the outdoor recreation industry, the alliance has recently been more active than the Society in promoting its vision of the trail. During the summer of 1997, the group initiated a program called "Uniting Along the Divide" to divide the trail into 31 100-mi segments with the use of GPS and ArcView. At the time of this writing, the path to be taken by the official route was unclear.

The two nonprofits have different visions of the trail and have not yet collaborated to identify a common solution. The alliance seeks well-established trail heads and tends to concur with federal routing decisions, while the Society seeks a less commercial trail that minimizes hiking along roads.

Status of Continental Divide Trail

An unofficial Continental Divide trail existed from Colorado to Montana at the time the trail was formally designated. In these states, it was possible to string together preexisting trails and roads to create a trail that follows the physical divide fairly closely, and the trail is nearing completion in these states. In New Mexico, however, no official route has been designated through long stretches of the Divide. Despite the absence of an established route, small groups of "thru-hikers" cross the state every year largely on paved highways and dirt roads—restrictions that diminish the wilderness experience considerably. To place the trail on private land, the owner must either grant an easement or sell a right of way. In New Mexico, neither course has proven successful, resulting in only 33 percent of the trail being designated, according to 1996 CDTA estimates.

Unfortunately, the enabling legislation was initially a stumbling block to trail construction. Section 551 of PL 95-625 specifically bars "any expenditure of funds for the acquisition of any lands, or interests in land, for the Continental Divide Trail." In New Mexico this policy prevented construction of a trail that closely follows the physical divide. Because no easements can be purchased, the BLM and the USFS have been forced to align the trail with paved state highways and dirt roads in areas where the divide crosses private land.

Section 551 of PL 95-625 was amended on March 5, 1980, as PL 96-199 to permit "the expenditure of funds for acquisition of lands or interests in land for the Continental Divide Trail within the exterior boundaries of existing Federal areas," and to allow federal matching grants to be used by state or local governments for trail purposes. With the use of high-resolution land use/ownership data, optimal trail routes can be better planned. If small sections of private land lie in an otherwise ideal corridor, communities can apply for federal matching grants to acquire land for the trail easement. Framework data supply federal land managers with information necessary to locate these problematic parcels.

However, the amendment refers only to lands within federal boundaries. BLM holdings do not have official boundaries and their holdings frequently change; only national parks, monuments, and forests have official boundaries. Therefore, the amendment does not include the main federal landowner in the state, the BLM, and it applies only if there is a willing seller. As a result, the amendment has gone largely unused. With higher resolution land ownership data in a GIS, the need for placing the trail within national forest or national park jurisdictions could be more easily accommodated. It is clear that PL 95-625, as amended into PL 96-199, contains the legal motivation for a GIS solution using high-resolution framework data.

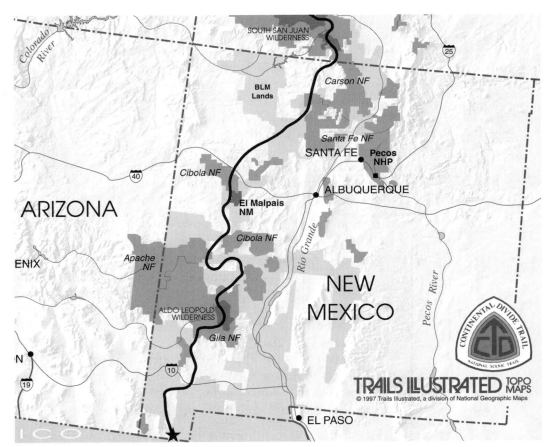

Current Continental Divide Trail in New Mexico. Courtesy of Trails Illustrated Topo Maps.

Status of Trail in New Mexico

An official trail complete with signage exists in certain areas of New Mexico. In the Gila National Forest and Aldo Leopold Wilderness Area, for example, it was possible to use preexisting trails as the Continental Divide Trail. In 1996, preexisting dirt roads and trails in the El Malpais National Monument/Conservation Area were designated with signage as the Continental Divide trail. In this general area of the state the following three segments continue to be embroiled in controversy: (1) the segment between the Gila National Forest and the U.S.-Mexico border in the "boot heel" of the state; (2) the segment between Gila and El Malpais; and (3) the segment north of Cuba, New Mexico. In this region, the trail would pass through either the Jicarilla Apache Indian Reservation or USFS land in the Carson National Forest, which in turn is part of the Tierra Amarilla Hispanic land grant. Numerous small parcels of private land also inhibit trail construction in other areas of the state.

Issues in New Mexico

Efforts to designate the Continental Divide trail in New Mexico touch on some of the state's most sensitive social and cultural issues. The main issues concern sovereignty of Indian reservations and Hispanic land grants, as well as local concerns about private property rights and opposition to federal control of public lands. The physical continental divide passes through two Indian reservations: the Ramah Navajo Indian Reservation just west of El Malpais National Monument, and the Jicarilla Apache Indian Reservation northwest of Cuba. The Ramah Navajo are adamantly opposed to having the official route pass through their land. This resistance led trail planners to place the route on BLM lands in the El Malpais National Monument/Conservation Area.

The Jicarilla Apache position is less clear. According to the USFS, the Jicarilla rejected a proposal to route the trail through their reservation, but the tribe's president seems willing to negotiate access for individual hikers. At present, the USFS has routed the trail through the Carson National Forest west of Tierra Amarilla. This plan is encountering resistance from local Hispanics sensitive about their right to privacy. For these reasons, the trail from Cuba to the Colorado border is further from completion than any other section.

The segment located in the state's "boot heel" is largely owned by the Animas Foundation, a nonprofit environmental conservation group. The group opposes passage of the trail through the Animas Mountains because of liability concerns. Drug trafficking is prevalent in the area, and there are concerns that hikers may run across bands of smugglers who may be armed and dangerous.

Solution

Locating the final route of the Continental Divide Trail will require fine-resolution framework data for land use and ownership, topography, roads, and trails. A prototype database has been developed in an attempt to identify the best trail routes. GIS offers an ideal solution to the problems surrounding trail location. When topography can be viewed in the context of land management responsibility and the transportation network, possible corridors for a trail are far easier to visualize. The database, more accurate than paper maps, can grow with the trail instead of being continually redrawn to update features. Even when the trail is in place, the GIS database would be useful for maintaining and expanding the trail. The prototype database will mark the first time that data from the USFS, BLM, NPS, and CDTS will be merged into a single database.

Identifying a route for the trail in the area north of Pie Town, New Mexico, illustrates the value of geospatially corrected, high resolution land ownership data. In this checkerboard region near the physical divide, the BLM proposes to route the trail along state highways. The CDTS, however, proposes to route the trail through USFS land to a checkerboard area of BLM, state, and private land just north of the forest, using common corners to remain on the BLM and state parcels until it emerges on BLM land near the headwaters of Cebolla Canyon. With sufficiently detailed data on land ownership, state roads, and topography, finding an appropriate route for the trail becomes a relatively simple task.

Thus far, all parties have been working independently—and sometimes at cross purposes. Surprisingly, there has been very little effort to ensure that the route proceeds along the common boundaries of the two federal agencies' holdings. In addition, there has been little or no communication between the CDTS and CDTA to reach a compromise. Therefore, a spatial database combining the work of all interested parties would be a valuable management tool. When complete, the USFS, BLM, NPS, CDTS, and CDTA will be provided with copies of the database.

Database Prototype

The prototype database under development contains the following major layers of data in ARC/INFO: topography (including the physical divide), existing roads, land ownership, land use, and proposed routes. All data were obtained courtesy of the New Mexico Resource Geographic Information System (RGIS), with the exception of the trail routes. The topography (1:100,000 scale) was created by the National Mapping Division of the U.S. Geological Survey (USGS).

The land cover data were in the form of 1:250,000 quadrangles. Quadrangles were joined to provide a contiguous coverage of the western part of the state. The categories of land use were reclassified to best represent the issues (forest, range, agricultural, urban land, wetlands, and bodies of water).

Land ownership data at a scale of 1:100,000 were obtained from the BLM. These data were joined and reclassified into several categories to emphasize major federal landholdings versus private landowners (BLM, USFS, NPS, private, Native American, state, and military).

The path of the Continental Divide through New Mexico was determined by a watershed boundary coverage at a scale of 1:500,000, also created by the USGS. The roads coverage was created by the Earth Data Analysis Center (EDAC) at the University of New Mexico using a Trimble GPS unit at a scale of 1:100,000. The proposed routes were digitized from maps provided by the local USFS, BLM, NPS, and the Continental Divide Trail Society.

Water sources were obtained from the BLM office in Albuquerque, New Mexico. The BLM data were recorded on USGS 7.5-minute maps at a scale of 1:24,000. The sources have been field checked with a Trimble Basic Plus GPS unit. Field notes were recorded on land ownership, as well as accessibility, quality, and seasonal reliability (when possible) of water sources. These data are to be incorporated into the database in the form of an attribute table.

Conclusions

The BLM, USFS, and the Society have all proposed viable routes. It appears, however, that the best trail would be a combination of federal agency and Society plans. In many sections there are no alternatives to routing the trail on roads. Paved roads, however, should be avoided. As Jim Wolf has pointed out in the Society's literature, "The National Trail System Act declares that national scenic trails should provide for maximum outdoor recreation potential and for the conservation and enjoyment of the nationally significant scenic, historic, natural, or cultural qualities of the areas through which such trails may pass. Locating the trail along heavily traveled roads would fail to honor these policies." The Society has proposed several viable alternatives to placing the trail on state highways that warrant further consideration by federal agencies. Hopefully, the prototype database will provide the necessary information to help create a Continental Divide National Scenic Trail in New Mexico that maximizes scenic beauty along with the wilderness experience.

Cartographic Support for Managing Washington State's Aquatic Resources

B.N. Dahlman and E.L. Lanzer,
Washington State Department of Natural Resources

Research Management Requirement

The Washington State Department of Natural Resources (DNR) manages more than 5 million acres of public land. Of the total, state forest lands account for 2.1 million acres; agricultural, approximately 1 million acres; and aquatic lands, 2.1 million acres. The agency's primary missions are to generate revenue for state trust land beneficiaries and provide natural resource protection and public services. The department has used ARC/INFO for more than 13 years and conducts forest land management in an enterprise GIS environment.

The department's Aquatic Resources Division manages state-owned aquatic lands, which include permanently submerged bedlands in navigable fresh and salt water, shorelands along navigable freshwater lakes and streams, and salt water tidelands. Resource management goals encompass promotion and enhancement of public access to state aquatic lands, maintaining commerce and navigation, ensuring environmental protection, and sustainable usage of aquatic renewable resources. The division is responsible for generating revenue from such lands in a manner consistent with these goals.

The Aquatic Resources Division is working toward using GIS in day-to-day management of the state's aquatic lands. A major benefit of the agency's substantial investment in GIS is that all of the Aquatic Resources Division's GIS software and hardware are maintained and supported by the department's Information Technology Division. This allows the Aquatic Resources Division's GIS team to focus its efforts exclusively on geospatial data development and upkeep, data analysis, and product generation. The Division's GIS team supports the following aquatic resource management programs: Ownership Records, Shellfish Management, Nearshore Habitat Inventory, and Spartina Control.

Aquatic Ownership Records Program

Developing output products for the division's Ownership Records program has presented some interesting cartographic problems. Many previous ownership exhibits used in courtrooms or other formal hearings consisted of enlarged aerial photographs with graphic line tape delineating ownership. Producing these exhibits proved to be labor intensive, and ultimately lacked the necessary positional accuracy.

In recent years, accurate aquatic ownership information has become available in ARC/INFO or AutoCAD digital format. Surveyed parcel data are converted from AutoCAD to ARC/INFO data sets. As the digital database is created, producing ownership exhibits directly from the GIS system becomes increasingly feasible. At the same time, the DNR has converted its orthophoto production from a hardcopy based system to a softcopy system. Digital grayscale orthophoto images are available for most tideland areas. Each image covers a single township at a 3-ft spatial resolution. The resulting tagged image file format (.*tif*) files average 225 Mb each. To generate exhibits of larger areas (i.e., covering more than four townships), the data must be loaded to a temporary large file system.

Halos allow custom line symbols to be visible over the highly variable tones characteristic of a digital orthophoto image.

Ownership boundary information is the focal point of any aquatic ownership exhibit. Unfortunately, the default line symbols in ARCPLOT could not be easily distinguished when overlaid on grayscale orthophotos. Both light and dark lines tended to blend into the highly variable tones on the digital image background. To provide exhibit users with easily distinguishable ownership boundaries, a custom line symbol set was created that included several color and pattern lines surrounded by a white "halo." The halo effect allows the line to be visible over any background value when opaque plotting options are used. As a final processing step, ARCPRESS was used to calibrate raster display values separately from vector display values to provide greater foreground/background distinction on the output.

Shellfish Management Program

The Shellfish Management program is responsible for managing the state's geoduck (pronounced GOOEY-duck) clam resources. Geoduck clams can be commercially harvested in subtidal state-owned lands, but harvesting must occur at least 200 yards from shore and from 18 to 70 ft below mean low water. The program sells resource harvest rights on a bed-by-bed basis in a manner designed to ensure a sustainable population. As such, it is responsible for defining the boundaries of harvest areas, as well as monitoring and enforcing harvest activities, including the quantity harvested.

The GIS team supports harvest planning activities by assisting in mapping geoduck beds. Differentially corrected global positioning system (DGPS) data and sonar depth finders are used to delineate beds. The DGPS data are converted into an ARC/INFO data set, and maps showing bed boundaries are produced for subsequent underwater resource surveys (see the following figure). The state's Department of Fish and Wildlife sends scuba divers down to the bed to establish clam density per acre and average clam size in order to determine harvest quotas. Rights to harvest the beds are then auctioned to the highest bidding commercial harvester. GIS products are also used in harvest planning activities, such as auction documentation and shoreline management permits, and in the field during pre-harvest tract layout and active harvest enforcement.

The impact of the multipurpose nature of cartographic products for the Shellfish Management program can be seen in several map elements and mapping activities. The agency's enterprise GIS database is stored in NAD 27. However, most "navigational grade" DGPS operate in NAD 83. Utilizing ARCPLOT's neatline functions, the output maps for fieldwork are reprojected to NAD 83. In addition, navigational grade DGPS output in degrees and decimal minutes. The commercial geoduck beds are mapped with "resource grade" DGPS that output degrees, minutes, and decimal seconds. To avoid confusion during field operations, all output maps represent longitude and latitude as degrees and decimal minutes. ARCPLOT's neatline functions easily facilitate changing output formats.

In addition to using the features contained in the cartographic software itself, performing fieldwork and assisting in commercial bed mapping has been invaluable for the GIS team. The staff has learned more about geoduck bed data, field data collection conditions, and how the data are used in the field by program and Department of Fish and Wildlife divers. Shellfish harvest field workers who have never used a digitizing table, for example, have gained experience in conceptualizing the boat as a virtual "puck." Likewise, office staff have found it beneficial to observe how the boat's movements and DGPS collection intervals affect post-processing time and effort. This is clearly a case where more data are not always a good thing. ARCPLOT and ArcView are used to generate posters and graphic slides of geoduck resource information for use at shoreline management hearings (where harvest permits are granted), at harvest auctions, or as courtroom exhibits.

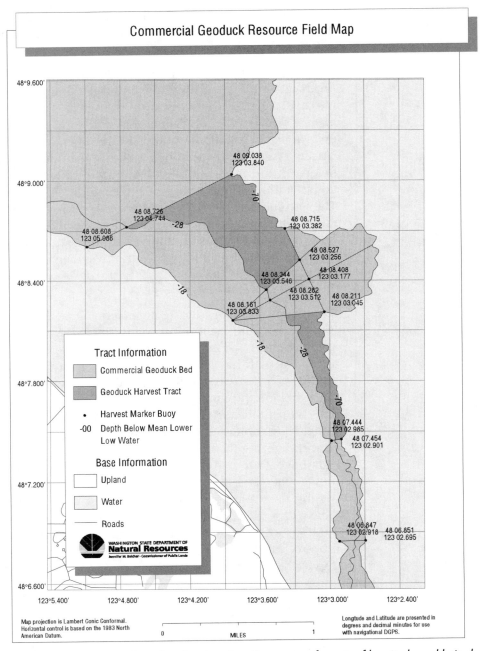

Commercial Geoduck Resource Field Map

Tract Information

- Commercial Geoduck Bed
- Geoduck Harvest Tract
- • Harvest Marker Buoy
- -00 Depth Below Mean Lower Low Water

Base Information

- Upland
- Water
- — Roads

WASHINGTON STATE DEPARTMENT OF
Natural Resources
Jennifer M. Belcher - Commissioner of Public Lands

Map projection is Lambert Conic Conformal. Horizontal control is based on the 1983 North American Datum.

Longtude and Latitude are presented in degrees and decimal minutes for use with navigational DGPS.

0   MILES   1

ARCPLOT's project and neatline functionality allow custom formats of longitude and latitude units and datum changes for field maps.

Nearshore Habitat Program

The Nearshore Habitat program uses GIS to maintain and analyze habitat data for Puget Sound. The program's nearshore habitat inventory contributes to information about the general health of Puget Sound. Comprehensive management of nearshore resources requires accurate and current information about resource location, abundance, and characteristics. These GIS data are used by the rest of the Aquatic Resources Division to determine if proposed land use activities will adversely impact important habitats.

Intertidal habitat data types include vegetation and substrate. The program receives multispectral image data from a Compact Airborne Spectrographic Imager (CASI) sensor. Imagery is collected in square 4m pixels and classified to identify vegetation types. Aerial photo interpretation and field work establish the substrate types and serve to determine accuracy of the classified CASI data.

The Aquatic Resources' GIS team has found that the technical and environmental complexity of its task makes coordination among the marine scientists, remote sensing, and GIS staff critical. Scientists in the field conduct portions of the GIS data capture process along with the GIS staff in a process helpful to both groups. Field data collection experience has helped GIS specialists to better understand the data set being created, and has led to changes in data models to better capture crucial complexities seen on the ground. Likewise, a hands-on GIS digitizing and data processing perspective has helped field scientists distinguish between traditional scientific research data collection (e.g., transects, quadrats, and population counts) and geographic mapping data collection requirements.

While in the field, marine scientists have become more aware of and conscientious about mapping issues, such as the minimum mapping unit, the scale at which the data will be digitized and displayed, and the generalization inherent in both manual digitizing and raster data capture techniques. For example, from a boat, a large vertical cliff face appears as a polygon with many important habitat attributes. However, when viewed on a planimetric map, the same cliff-side habitat polygon appears as a line and falls below the minimum mapping unit of the project. Overall, "cross training" between field and GIS staff has improved the program's ability to create, store, and display habitat mapping.

Cross training is also reflected in cartographic support for the program. GIS staff in the field noticed that managing large map sheets in an open boat—especially in the often-soggy Puget Sound weather—was clearly making one phase of field work cumbersome. The large map sheets were being used because they showed all of the project's several hundred field sites. The solution: center each site on an 8.5 x 11 in page, include all necessary base information, and print the maps on a color laser printer on waterproof paper.

In addition, a 1:24,000 series of nearshore habitat maps was designed and published showing four sets of data for each quadrangle. The linear nature of nearshore habitats and the fact that the substrate and vegetation components of habitats generally occupy the same geography made the habitats difficult and complex to represent with standard symbology. By using custom shade symbols, even the two most detailed data sets were displayed as if overlayed within one map window (see the next image). Dot symbol sizes for the vegetation were established by closely replicating CASI pixel sizes. The result ensures that each polygon will receive some significant portion of a dot, allows dot symbols to be easily identified by color, and metaphorically represents the discontinuous nature of vegetation on top of the continuous substrate.

The map series has been published on CD-ROM for a portion of Washington state. This method of data and map product distribution may well be expanded as the inventory for more geographic areas is completed. Other cartographic products include presentation graphics for use as slides or in publications that are vital to meeting the program's public education and research mandate.

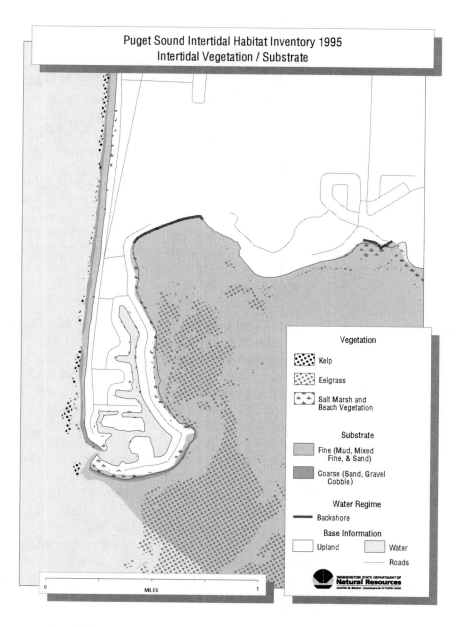

Nearshore habitat map.

Spartina Control Program

Spartina is a non-native invasive weed artificially introduced to the Pacific Northwest. As the plant spreads, it replaces native vegetation and covers mud flats. Such mud flats are important to shorebirds and critical to aquaculture, a traditional and important part of affected regions' local economy. The Department of Aquatic Resources' Spartina Control program uses ARC/INFO and ArcView to store and analyze information about state-owned lands infested with spartina, and help coordinate weed control efforts undertaken by federal, state, and local agencies. Output generated by ARC/INFO for this program includes standard 1:24,000 and 1:12,000 field map series, poster-size graphic maps, and 35mm slides used at public hearings and legislative presentations. These products have been critical working tools for planning ground and aerial control activities, and are beginning to illustrate the effect of control efforts as the program's monitoring phase proceeds.

Conclusions

The division's GIS team and field staff have benefited greatly from participating in field data collection and helping to convert field data to a geospatial data set, respectively. Developing a greater understanding of the data collected and how resource managers use them has helped the GIS team produce better data sets and more functional map products. The Aquatic Resources GIS team has found it useful to create custom symbols to resolve specific cartographic issues. Consideration for cartographic output during data design has helped reduce the creation of derivative data sets sometimes required to achieve a specific cartographic effect. Examples of cartographic consideration in data design include label positioning within polygons (while digitizing) to facilitate proper annotation placement, ensuring that all arcs point in a predetermined direction to facilitate line offsets, and adding a symbology code item (or RELATE) for direct polygon shading without a look-up table or numerous reselections. ARCPLOT and additional tools such as ARCPRESS provide a useful set of capabilities for effective cartographic production. However, the ultimate responsibility for using these tools to create maps—rather than mere graphics—requires having the ability to apply essential cartographic design principles.

Multi-scale Resource Data

R.G. Congalton, University of New Hampshire

Natural resource managers have long used spatial data in support of their work (e.g., Spurr 1948, Colwell 1960, and Thompson 1979). Examples of how spatial data were traditionally used include simply making a sketch or a map of what was observed, making some annotations on top of another map such as a U.S. Geological Survey 7-1/2 minute quadrangle, and/or delineating boundaries or other information on an aerial photograph or satellite image. The needs for creating these maps are as numerous as the maps themselves. They include forest inventory, forest health, wildlife habitat, soil characterization, rangeland monitoring, wetland delineation, conservation management, timber harvesting, land use zoning, recreation identification, hazard area demarcation, geologic exploration, and ecosystem restoration to name just a few.

Most maps created prior to the GIS era are only loosely tied to the ground. In other words, although each of such maps is a representation of some area on the Earth, most have no means of exactly locating any point on the map to the ground. The next illustration is a property boundary map and is a classic example of a map of this type. Note that bearings and distances are given and yet there is no way to place this map exactly on the Earth because no point of known ground location is provided. In many instances, some type of marker was described as the starting point of the survey (e.g., new marker, or existing fencepost, or even a tree or boulder). However, these markers have long since vanished and now it is very difficult to know where exactly one is on the ground.

Typical example of map with no known ground locations.

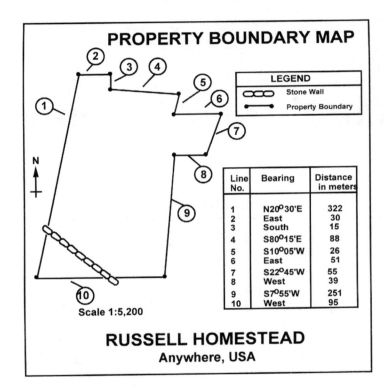

PROPERTY BOUNDARY MAP

LEGEND
- Stone Wall
- Property Boundary

Line No.	Bearing	Distance in meters
1	N20°30'E	322
2	East	30
3	South	15
4	S80°15'E	88
5	S10°05'W	26
6	East	51
7	S22°45'W	55
8	West	39
9	S7°55'W	251
10	West	95

Scale 1:5,200

RUSSELL HOMESTEAD
Anywhere, USA

For most resource applications, not knowing exactly where you are on the ground has not been a serious problem. After all, a map is understood to be simply an abstraction or representation of what is on the ground. A map does not have to be perfect; it simply has to be useful for a given objective. For example, the next two illustrations present a timber type map and a soil type map, respectively. Again, these maps have no known ground locations to tie them to a precise location on the Earth. Both have been generated using the property boundary map outline. Obviously, one could go to the ground with this map and it would help in navigating and aid in resource decision making. Questions about the exact location of a certain feature (e.g., tree, inventory plot, soil pit, or boundary between vegetation types) were not considered economically or technically realistic before GIS and GPS (global positioning system) technology.

Typical forest type map.

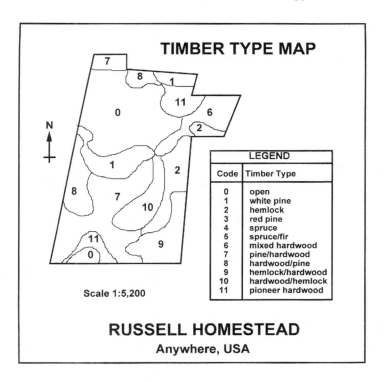

Typical soil type map.

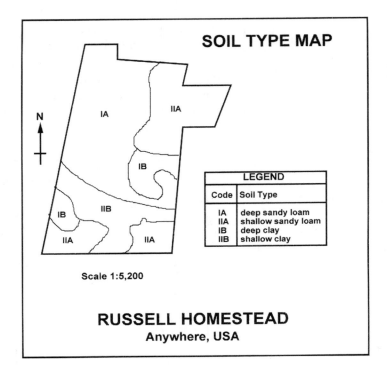

SOIL TYPE MAP

N

IA

IIA

IB

IIB

IB

IIA

IIA

LEGEND	
Code	Soil Type
IA	deep sandy loam
IIA	shallow sandy loam
IB	deep clay
IIB	shallow clay

Scale 1:5,200

RUSSELL HOMESTEAD
Anywhere, USA

With the advent of GIS technology, a major paradigm shift occurred. Not only is it possible to have very detailed maps and quick analysis of multiple map layers, but it is possible by using GPS on the ground to know the location of anything and to incorporate such locations in a map.

As a result of this increased mapping capability using GIS, two new and very significant mapping issues arose almost immediately. This first issue is map scale. Historically, maps have been viewed at the scale in which they were created. This is good and right since only the appropriate details at that scale have been recorded on the map. However, if the map is input into a GIS, it can be reproduced at virtually any scale. Big problems can result. The second issue is one of accuracy or error. As already mentioned, maps are general abstractions of the ground and embody certain expectations and obvious limitations. However, computerized maps in many colors with detailed legends and labels inherit a whole new set of expectations and less obvious limitations. These new and more rigorous expectations are not necessarily warranted in many circumstances.

Map Scale

A major concern in collecting and using spatial data for natural resource management, or any mapping purpose, is scale. Most of the GIS data available are collected at the scale of common base maps. In the United States, the largest common scale at which data are collected is 1:24,000; the scale of the U.S. Geological Survey 7-1/2 minute quadrangles. Scales of 1:50,000 are more common outside the United States. Smaller scale data are collected at 1:100,000 and 1:250,000 scales. In natural resource management, maps have often been generated at 1:15,840 (4 inches to the mile) and 1:12,000 because aerial photography is often acquired at these scales.

In the early years of GIS usage, most effort was devoted to collecting and building GIS databases. Data at these common scales (i.e., 1:24,000 in the U.S. and 1:50,000 internationally) were sufficient to conduct many large area and regional projects and took less effort to collect than more detailed information required at larger scales (e.g., 1:6,000). Conducting a GIS analysis of deer habitat in New England or even New Hampshire could be conducted using the smaller, common scale databases. Analyzing old growth forest patterns in the Pacific Northwest could also be performed with these data.

However, with the advent of GPS (global positioning systems), the development of improved computer speed and disk storage, and more sophisticated GIS software, these scales are no longer adequate for many applications. Larger scale maps and databases are required for more detailed mapping for local applications. While it is possible to change the scale of any map in a GIS, making the scale larger is not advised because the data were collected at a smaller scale and therefore, important details may be missing. (See section below on error in GIS data.) For example, a map of the state of New York collected at a scale of 1:50,000 will contain all major roads and highways, but not all the side streets. No matter how much the scale is enlarged, the side streets will not appear because they are not a part of the map. Such a map is perfectly acceptable for planning a trip across the state, but will be inadequate for attempting to locate someone's home in a small town.

Many other examples demonstrate this multi-scale or multi-resolution characteristic of GIS data sets. Two examples, digital elevation data and soils data, will be discussed to further illustrate this point.

Digital elevation models (DEMs), also called digital terrain models (DTMs), are digital files containing a grid pattern of point elevations that can be used to simulate an area's topography. These data are useful for generating three-dimensional information about land area. Examples of this type of information include slope, aspect, volume, and surface profiles. The U.S. Geological Survey provides digital elevation model data at two resolutions. A complete database for the entire United States was derived from the National Imagery and Mapping Agency (NIMA, formerly Defense Mapping Agency) data for 1° latitude x 2° longitude (1:250,000 scale) maps. The grid interval for these data is 3 arc-seconds or approximately every 100 meters. In addition to the NIMA data, the USGS also has an incomplete coverage of the United States using a grid interval of 30 meters. These data sets, commonly known as DEM data, are available in 7.5 minute blocks and are of much higher resolution than the NIMA data.

Using the above two sources of digital elevation data sources will result in very different maps. The NIMA data samples only every 100 meters, and will therefore be more generalized and contain less detail than the USGS data. The specific objective of the project will dictate which of these elevation data sets to use. For large areas (i.e., multiple states), the NIMA should provide sufficient detail. For smaller regions, the USGS data will provide the required extra detail. Neither of these elevation data sets may be appropriate for very local applications. In this case, a new digital elevation model with finer resolution and including samples much closer together than 30 meters (i.e., ≈ 1/2 arc second) may be required.

The same multi-scale/multi-resolution issues exist for the soils databases produced in the United States by the Natural Resource Conservation Service (NRCS). The most detailed (i.e., largest scale/finest resolution) soils maps are the Soil Survey Geographic Database (SSURGO) and are generated using standard county soil survey techniques at scales between 1:15,000 and 1:24,000. STATGO soils maps are then created by generalizing the SSURGO maps to a scale of 1:250,000. When SSURGO data are not available to be generalized, STATGO maps are then produced by combining data on geology, topography, vegetation, and climate in conjunction with remotely sensed data.

Selecting soil data having the appropriate scale and resolution is conditioned by the intended application. Failure to be informed about the scale/resolution at which the map data were collected can lead to serious errors in natural resource analysis.

Map Error

In order to make effective use of GIS technology, understanding the errors associated with spatial information is important (Goodchild and Gopal 1989). This knowledge is critical for GIS users and suppliers of spatial information (i.e., data layers) used in the GIS. It is also important to establish data standards and methods for documenting GIS data, called *metadata*.

Sources of Error

Errors associated with spatial information can be divided into the following categories: (1) user errors, (2) measurement/data errors, and (3) processing errors (Burrough 1986). User errors are probably the most obvious and are more directly under the user's control. Measurement/data errors deal with spatial information variability and the corresponding accuracy with which the data were acquired. Finally, processing errors are inherent in the techniques used to input, access, and manipulate spatial information. Error sources are summarized in the following table along with measures of magnitude and difficulty in dealing with the effects of error. Combining these two factors allows one to establish priorities for avoiding or reducing the impact of error.

Error source	Error contribution potential	Error control difficulty	Error index	Error priority
USER				
Age	2	3	6	6
Scale	3	4	12	11
Coverage/extent	1	2	2	1
Indirect/derived layer	2	4	8	7
MEASUREMENT/DATA				
Instrument error	2	4	8	7
Field error	1	3	3	4
Natural variation	1	5	5	5
PROCESSING				
Precision	1	2	2	1
Interpolation	2	4	8	7
Generalization	3	3	9	10
Conversion	3	4	12	11
Digitization	1	2	2	1

Error contribution potential–Relative potential for this source as contributing factor to the total error (1 = low, 2 = medium, and 3 = high).

Error control difficulty–Given current knowledge of this source, difficulty in controlling the error contribution (1 = not very difficult to 5 = very difficult).

Error index–Index representing the combination of error potential and error difficulty.

Error priority–Order in which methods should be implemented to understand, control, reduce, and/or report the error due to this source based on the error index.

User Error

User errors include data age, scale, coverage, and relevance. Errors result when obsolete data are used in lieu of more current information. A good example of this problem is using old aerial photography because newer photography is not accessible or is too expensive. Expediency of source materials can lead to serious errors in subsequent analysis.

Error also results when data of the wrong scale are applied to meet a particular objective. This situation is especially dangerous when small-scale data are used to meet the objectives of a larger-scale project. An example of this problem is using a statewide soils map to obtain soils information about a particular county or using digital elevation data from a 1:250,000 U.S. Geological Survey map sheet for mapping on a 7.5-minute quadrangle. In both cases, the source data are of insufficient detail to meet the required objectives.

In addition, errors result from using data sources that do not completely cover the area of interest. Many times only partial areas have been covered by the latest data and one is forced into the dilemma of choosing between full coverage using obsolete data or newer data with only partial coverage.

Finally, errors result from using indirect or derived data layers as input into the GIS. At issue here is the relevance of the derived data layer. An indirect layer has been derived from primary information. A good example of derived data is a vegetation type map generated from satellite data or aerial photography. Errors are introduced by creating the derived layer. If the process is rigorous, the derived layer may be more than adequate for the desired purpose. However, even in best case scenarios, a vegetation map generated from remotely sensed data will not be 100% accurate.

Measurement/Data Error

Included within measurement/data error are errors associated with variations in the data set. Among these variations are instrument error, field error, and natural variation. Instrument error is simply a measure of the limitations and/or the quality of the instrument being used to collect the information. In like manner, field error is a measure of the limitations and/or the skill of the person collecting the data. Natural variation is a condition of nature. As much as one might wish, nature is a continuum; it cannot be compartmentalized. It is extremely difficult to distinguish between two mixed stands of trees, one of which is 45% conifer/55% hardwood (i.e., hardwood-conifer) and the other, 55% conifer/45% hardwood (i.e., conifer-hardwood). Therefore, although we must live with this error, we must also make every effort to account for it in developing spatial information.

Measurement/data errors are usually quantified in terms of positional accuracy or accuracy of content. In other words, "Am I where I think I am, and if so, are my surroundings what they should be?"

Processing Error

Finally, processing errors include factors such as precision, interpolation, generalization, data conversion, digitization, and other methodological operations. An often overlooked fact is that computers are designed for only a certain level of precision. Going beyond that limit results in round-off error. Almost every introductory course in computer science discusses precision and the use of significant digits, yet often these factors are not considered when processing data. The most recent example of forgetting this issue came about when the Pentium computer chip was first released. There was a bug in the chip that rounded off large numbers and caused significant errors when making certain calculations.

In cases of interpolation, extrapolation, and generalization, information is derived about an area from a series of sample points. *Interpolation* is best described as an educated guess. It is predicting information about an unknown point from surrounding points. *Extrapolation,* on the other hand, is simply a guess. It is predicting information about an unknown point from points that are far away. *Generalization* is the elimination of certain known redundant sample points. DEMs are a good example of all three. In some places, interpolation is used to increase the number of sample points. In other cases where the density of points is insufficient, extrapolation is necessary to increase the sample points. Finally, where too many data points exist, generalization may be employed. Together, these three operations are used to obtain the distribution of data points needed to create a DEM.

Next, the way the data are entered into the GIS is subject to error. The data can either be scanned or digitized but in both cases errors can occur. Some of this error is due to instrument limitations while some is the result of human error.

The ways in which the data are stored and used in a GIS may introduce errors. This can occur, for example, when converting between vector and raster data (Congalton 1997). Digital satellite data are recorded in pixel format, also known as raster or grid format. Other information in the GIS may be recorded in vector or polygon format. Therefore, in order to merge these data sets, the vector data may need to be converted to raster or vice versa. Errors could result from making smooth polygons into grid cells or making grid cells into smooth polygons.

In addition to the above processing errors, problems arise when data layers are combined to perform analyses. Consider the problems associated with overlaying data layers, boundaries, and registration. These factors and others result in error in the processing stage of using spatial information.

Measuring Error

Once sources of error associated with spatial data are understood, it is also important to learn how to measure them. Three measurement considerations come to mind. First, sampling must be performed to obtain information about the error. Second, statistical analyses are required to test hypotheses about the error. Finally, the patterns of error should be explored. None of these should be undertaken until something is known about the characteristics of the spatial data.

Data characteristics, in the statistical sense, refer to whether or not the data are continuous or discrete and exactly how the data are distributed (i.e. normal, binomial, multinomial, etc.). A continuous variable is one that can take on any value within any observed range, such as the height of a tree or the width of a building. Discrete variables, on the other hand, have only specific values; examples are the number of leaves on a tree or the number of bricks in a building. Many common statistical techniques assume that the data of interest are continuous and distributed normally. Herein lies the problem–most of us have learned something about normal theory statistics and have been informed that they can apply to any situation. Spatial data, however, are rarely continuous and normally distributed. In fact, most spatial data (e.g., all remotely sensed data) are discrete or categorical in nature. Even continuous spatial data such as elevations are frequently represented in discrete categories (e.g., 10m contour intervals). It is important to understand the characteristics of the spatial data before further analyses are performed.

Once the characteristics of the data have been defined, the information can be incorporated into the sampling scheme and statistical procedures used to assess error. *Spatial autocorrelation* analysis is helpful in choosing the proper sampling scheme (Congalton 1988). This analysis tests the influence or effect that a certain quality or characteristic at a given location has on that same quality at neighboring locations. In other words, given a group of mutually exclusive units in a two-dimensional plane, if the presence, absence, or degree of a certain characteristic affects the presence, absence, or degree of the same characteristic in neighboring units, then the phenomenon is said to exhibit spatial autocorrelation. Spatial autocorrelation then dictates the most appropriate sampling scheme to use in order to assess the error. The common schemes such as *simple random sampling* and *stratified random sampling* are always appropriate. Other schemes such as *cluster sampling* and *systematic sampling* should be evaluated in light of the autocorrelation results. For example, if the error is clustered, use of a systematic or cluster sampling approach may severely misrepresent the error in the data.

Finally, statistical analyses must be determined based on the characteristics of the data. If the data are discrete and/or not normally distributed, then techniques other than parametric statistics should be considered. While in many instances the central limit theorem and large sample sizes permit the use of normal theory statistics, one should also consider nonparametric or discrete multivariate procedures.

Assessing errors in spatial data is an important yet difficult task. There is much work yet to be done in this area. As more and more decisions are made using spatial data, it will become increasingly important to know where each data layer originated and the accuracy of each layer. A simple example emphasizes this concept quite strongly (see the next illustration). If three data layers–each with an accuracy of ninety percent–are overlayed to make a fourth layer, the best possible accuracy of this layer is ninety percent and the worst possible accuracy is seventy three percent. The best possible accuracy is achieved if the errors in the three data layers are completely correlated (i.e., in exactly the same place in each data layer). The worst possible accuracy is achieved if the errors in the three data layers are completely uncorrelated or independent. In this case, the accuracy is computed by multiplying the accuracy of the three data layers (i.e., .90 x .90 x .90 = 73%). The expected accuracy lies somewhere between these two values depending on the correlation between the three data layers. In many analyses, more than three data layers are used and most data layers are not ninety percent correct.

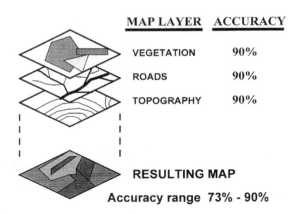

Example of overlay error.

MAP LAYER	ACCURACY
VEGETATION	90%
ROADS	90%
TOPOGRAPHY	90%

RESULTING MAP

Accuracy range 73% - 90%

Conclusions

GIS has certainly changed the way natural resource managers think about and use spatial data. Maps and remotely sensed data have long been a part of the manager's tool kit. It would be misleading to say that GIS has made the job easier. However, GIS permits detailed analysis and planning that were impossible before this technology became available. Therefore, better decisions are possible and new applications are being developed every day. Questions about the accuracy of the data are more relevant today than ever before. With the ability to do more comes the added responsibility of doing it correctly and documenting what has been done.

References

Burrough, P.A. *Principles of Geographical Information Systems for Land Resources Assessment*. New York: Oxford University Press, 1986.

Colwell, R.N., ed. *Manual of Photographic Interpretation*. Washington, D.C.: American Society of Photogrammetry, 1960.

Congalton, R. "Using spatial autocorrelation analysis to explore errors in maps generated from remotely sensed data." *Photogrammetric Engineering and Remote Sensing* 54:5 (1988), 587-592.

Congalton, R. "Exploring and Evaluating the Consequences of Vector to Raster and Raster to Vector Conversion." *Photogrammetric Engineering and Remote Sensing* 63:4 (1997), 425-434.

Goodchild, M. and S. Gopal, eds. *The Accuracy of Spatial Databases*. Taylor and Francis, 1989.

Spurr, S.H. *Aerial Photographs in Forestry*. New York: The Ronald Press Company, 1948.

Thompson, M.M. *Maps for America, U.S. Geological Survey*. Washington, D.C.: Government Printing Office, 1979.

Modeling Conservation Priorities in Veracruz, Mexico

S. Egbert, A.T. Peterson, and K. Price, University of Kansas; V. Sánchez-Cordero, Universidad Nacional Autónoma de México

Resource Management Requirement

Throughout the world, designating conservation and protection areas (e.g., state and national parks, wildlife preserves, and wilderness areas) has been stimulated by a variety of goals. These goals, among many others, include historical significance, recreational use, scenic beauty, public accessibility, and protection of natural features, such as geologic formations, forests, wetlands, and watersheds. Because biodiversity considerations have generally not been taken into account, biodiversity preservation is rarely optimized in existing systems. While parks and preserves may afford protection to some species of conservation interest, other species may be left without protection. In addition, some protected areas may contain no species of conservation interest.

In recent years, numerous efforts have emerged to identify gaps in biodiversity protection, such as the research by Scott et al. (1996). One technique for producing maps of biodiversity gaps is to first map predicted distributions for species of interest based on known habitat characteristics. Maps of predicted distributions are then compared with existing protected areas to identify gaps. Modeling habitat for organisms of conservation interest gives planners and decision makers tools to assist them in efforts to optimize biodiversity preservation.

This case study focuses on species distribution modeling and analysis of existing and optimal preservation systems for bird and mammal species that are native to northeastern Mexico. Geographic distributions for 19 species were first predicted, and then gaps in their protection were identified. Finally, a possible solution was developed for enhancing protection of the species within the context of the current system of protected areas.

Methodology

Two types of data were used in the analysis: species distribution data and environmental attribute data in a raster GIS architecture. Species selected for inclusion in the study are native to eastern Mexico, with primary habitat in Veracruz. Eight mammal and 11 bird species were selected, based on maps published in *Mammals of North America* (Hall 1981) and *A Guide to Birds of Mexico and Northern Central America* (Howell and Webb 1995). Distributional data for the study species were taken from records in the scientific literature and from systematic collections. (See Peterson and Sánchez-Cordero 1994 and Prieto and Sánchez-Cordero 1993.) A total of 69 unique observation localities were identified for the 19 species (see the following table). Observations were georeferenced to topographic maps (1:250,000 scale) from the Instituto de Estadística, Geografía e Informática (INEGI).

Native birds and mammals used as basis for modeling conservation priorities in Veracruz, Mexico

Birds	Known points	Mammals	Known points
Amazona viridigenalis	4	Dasyprocta mexicana	4
Atlapetes apertus	8	Megadonthomys nelsoni	1
Campylopterus excellens	12	Microtus quasiater	7
Caprimulgus salvini	3	Othogeomys lanius	1
Cyanolyca nana	2	Peromyscus bullatus	1

Birds	Known points	Mammals	Known points
Dendrortyx barbatus	4	Peromyscus furvus	1
Doricha eliza	3	Spermophilus perotensis	4
Geothlypis flavovelata	1	Sorex macrodon	5
Geotrygon carrikeri	2		
Hylorchilus sumichrasti	1		
Rhodothraupis celaeno	5		

The following thematic geographic layers were used to define environmental variables.

- Climate (18 categories)

- Elevation (12 categories at 500m intervals)

- Total annual precipitation (9 categories)

- Soils (18 categories)

- Mean annual temperature (5 categories at 2°C intervals)

- Vegetation/land use (45 categories)

The six thematic layers were georectified to each other and combined to create a single multilayer image file. All processing was performed using the ERDAS Imagine image processing and raster GIS software package.

Once the geographic data layers were combined, the species distributional data points were used to "probe" the layers; that is, the values for each of the geographic layers were extracted for each of the observation points.

Probing the geographic data layers to extract parameters for modeling species distributions.

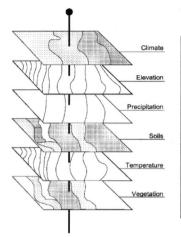

Sample Data Values for Geographic Layers				
Point	1	2	...	8
Climate	5	4	...	7
Elev	2	3	...	4
Precip	9	6	...	9
Soils	1	7	...	10
Temp	10	10	...	9
Veg	14	41	...	14

Using the range of values of each geographic layer for each species, a conceptual six-dimensional "box" (or parallelepiped) was constructed for each species. The limits of the parallelepiped represented the defining criteria for species distributions. This conceptual box was implemented spatially in ERDAS Imagine using the Spatial Modeler to select all pixels within the Veracruz study area that met the criteria defined by the limits of the parallelepiped.

Modeling species distributions. The six layers are "filtered" through a decision rule to produce a predicted distribution map. The rule shown is for the Plain-breasted brushfinch (Atlapetes apertus).

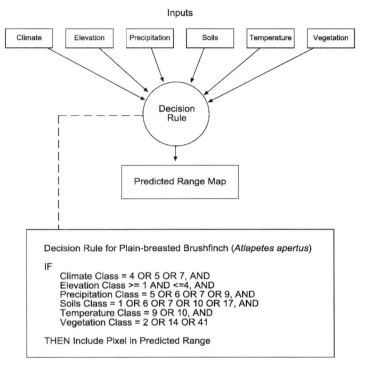

Inputs

| Climate | Elevation | Precipitation | Soils | Temperature | Vegetation |

Decision Rule

Predicted Range Map

Decision Rule for Plain-breasted Brushfinch (*Atlapetes apertus*)

IF
　　Climate Class = 4 OR 5 OR 7, AND
　　Elevation Class >= 1 AND <=4, AND
　　Precipitation Class = 5 OR 6 OR 7 OR 9, AND
　　Soils Class = 1 OR 6 OR 7 OR 10 OR 17, AND
　　Temperature Class = 9 OR 10, AND
　　Vegetation Class = 2 OR 14 OR 41

THEN Include Pixel in Predicted Range

Predicted range maps were produced for each species, visually examined for plausibility, and modified if necessary based on field knowledge of species distributions. Of the 19 predicted species distributions, six clearly were overestimated, based on published sources (Howell and Webb 1995) and field observations. To correct for obvious errors in the predicted distributions, areas of overestimation were set to zero in the GIS distribution maps.

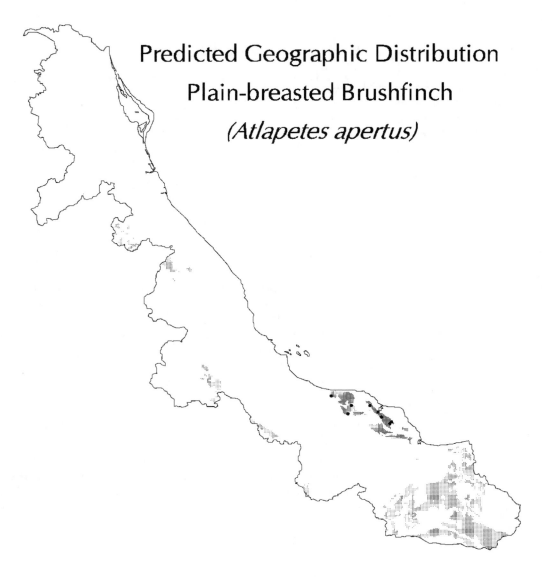

Predicted Geographic Distribution
Plain-breasted Brushfinch
(Atlapetes apertus)

Predicted distribution of Plain-breasted Brushfinch. Black dots indicate points of known presence, dark gray areas represent predicted locations, and light gray areas represent apparent overestimations.

In the final step, an optimal reserve system was identified based on the predicted distributions of the species. A reserve size of 100,000 ha was assumed; that is, a 20-pixel buffer was created around each predicted distributional pixel for each species, resulting in 19 buffered distributional maps. Imagine's Spatial Modeler was then used to sum the buffered species distributional maps, creating a map of species richness. The area with the highest species richness (seven species) constituted the highest conservation priority. Species present in this area were identified and removed from consideration in successive iterations. The species distributional maps for the remaining species were again summed, and the area with the highest species richness (five species) was identified as the next conservation priority. This was continued until the highest species richness value was one species.

For comparison, the existing protected areas in Veracruz were evaluated similarly. Each existing park or preserve was assumed to be 100,000 ha in size. (In actuality, all are much smaller.) In other words, a buffer of 100,000 ha was created around the central point of each protected area. Each of the protected areas was then evaluated with regard to species richness.

Results

The effectiveness of the current system for protecting the 19 native species was examined. There are currently 10 preserves in Veracruz, ranging in size from 49 to 55,900 ha. However, because boundary maps were not available, each was treated as if it were 100,000 ha in size. A simple raster GIS overlay operation was used to compare the buffered protected areas with the adjusted species richness maps. Of the 19 species examined, 12 are found in the current protected areas, while seven species are left unprotected.

Existing protected areas in Veracruz

Protected Area	Area (hectares)	Species protected
Cofre de Perote (PN)	11,700	4
Cañon del Río Blanco (PN)	55,900	0
Pico de Orizaba (PN)	19,750	0
El Morro de la Mancha	49	0
Los Tuxtlas	700	5
Sierra de Santa Martha	20,000	3
El Gavilán	unknown	2
Santa Gertrudis	925	2
San José de los Molinos	2,995	3
Volcán de San Martín	1,500	4
Note: PN indicates a national park.		

The next step was to identify an optimal preservation system for the 19 native species, while ignoring the current one. Using the approach described above, a system was generated that would protect the greatest number of endemic species in the smallest number of reserve areas. The analysis was able to include 16 of the species in four reserves and all 19 in seven reserves.

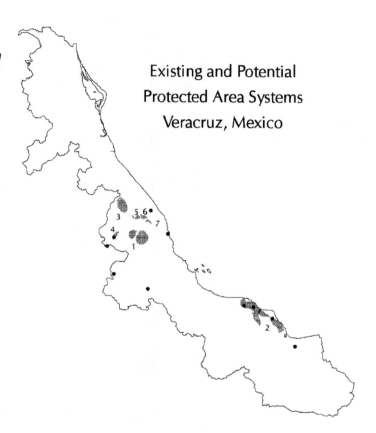

Existing and potential protected areas in Veracruz. Black dots indicate existing protected areas, and dark gray areas represent potential new protected areas. (Numbers show the order in which the areas were selected during the modeling procedure.) Numbers 1, 3, 5, 6, and 7 combined would provide additional protection needed for the 19 species in this study.

Existing and Potential Protected Area Systems Veracruz, Mexico

Clearly, biodiversity reserves created using optimization algorithms are more efficient at preserving biodiversity than the current system, which—as noted previously—was created with a variety of motives in mind. However, given the existence of a system of protected areas, it is intuitively unreasonable to expect that the current system would (or should) be scrapped in favor of an optimal reserve system. An alternative approach is to examine the potential reserves in the optimal system that would best complement the existing reserves by adding the unprotected species. By adding five of the potential protected areas from the optimal system to the current protected system, all 19 of the endemic species would be protected. Such a hybrid approach seems much more reasonable in real world conservation efforts, because it takes advantage of the existing protected area system and requires minimal new additions to the system to significantly expand biodiversity protection.

References

Hall, E.R. *The Mammals of North America*, 2nd ed. New York: Wiley, 1981.

Howell, S.N.G. and S. Webb. *A Guide to the Birds of Mexico and Northern Central America*. Oxford: Oxford University Press, 1995.

Peterson, A.T. and V. Sánchez-Cordero. "Un debate sobre la taxonomia en México: Nuevas ideas, nuevas metas, y un Estudio Biológico Nacional." *Boletín Academia de la Investigación Científica* 16 (1994).

Prieto, M. and V. Sánchez-Cordero. "Uso de un sistema de información geográfica: un caso de estudio mastofaunístico en Veracruz," pp. 455-463. In R.A. Medellín and G. Ceballos, eds., *Avances en el Estudio de los Mamíferos de México*. México, D.F.: Publicaciones Especiales de la Asociación Mexicana de Mastozoología, 1993.

Scott, J.M., T.H. Tear, and F.W. Davis, eds. *Gap Analysis: A Landscape Approach to Biodiversity Planning*. Bethesda, Maryland: American Society for Photogrammetry and Remote Sensing, 1996.

Development of a GIS Hydrography Database for Phosphorus Transport Modeling in the Lake Okeechobee Watershed

J. Zhang, L. Moore, W. Guan, A. Essex, G. Ritter, and L. Price,
South Florida Water Management District

Resource Management Requirement

As a result of increasing phosphorus (P) concentrations in Lake Okeechobee and associated water quality problems such as algae blooms, the South Florida Water Management District (SFWMD) has implemented numerous P management programs in the lake's watershed. These programs include water quality monitoring and modeling, research, regulation, and voluntary efforts directed at implementation and evaluation of best management practices (BMPs) (SFWMD 1993). Water quality monitoring and modeling efforts have focused on determining the effectiveness of BMPs in reducing P loads from agricultural runoff. One major modeling effort was the development of a basin-scale P transport model (Wagner et al. 1996, Zhang et al. 1996).

This P transport model is a modeling system that simulates runoff and P movement from upland fields to a basin outlet, and consists of several submodels. One submodel is the Chemicals, Runoff, Erosion from Agriculture Management Systems - Water Table (CREAMS-WT) (Heatwole et al. 1987, 1988), which is a field-scale model that simulates runoff and phosphorus loadings from individual fields. The second submodel is the Enhanced Stream Water Quality Model (QUAL2E) (Brown and Barnwell 1987) that was modified to simulate phosphorus transport through wetlands in addition to stream channels (Wagner et al. 1996). The third submodel is the Hydrologic Engineering Center-2 (HEC-2; see

U.S. Army Corps of Engineers 1990) that simulates channel hydraulics and provides hydraulic data required for input to QUAL2E.

Hydraulic data include discharge, flow depth, width, and velocity in both channel and wetland systems. These data are derived from stream characteristics and network topology using HEC-2 software. The stream characteristics and network topology data are stored in several formats (i.e., AutoCAD drawings, ARC/INFO coverages, and reference materials), making the development of input files for the P transport model a tedious process.

To accelerate the process of preparing hydraulic data for the P transport model, a GIS hydrography database was developed. The study objectives follow: (1) develop a GIS coverage (data layer in ARC/INFO) including the hydrography and associated attributes for major, secondary, and tertiary streams; (2) add stream names, vegetation types, and Manning's roughness coefficients to coverage attributes; (3) view stream geometry (cross-section data) in a GIS environment; and (4) discretize the stream system by reach numbers using GIS algorithms based on the stream network topology and associated attributes. A stream reach is defined as a section with a unique combination of geometry and vegetation.

Basin Description

The Lake Okeechobee watershed is approximately 6,000 km^2 in size and contains 33 basins (see next illustration). Among the 33 basins, the Taylor Creek-Nubbin Slough basin (S-191) and the Lower Kissimmee River basins (S-154, S-65D and S-65E), located mainly in Okeechobee County, have been identified as key basins that account for the majority of the excessive P loads to the lake (SFWMD 1997). A GIS hydrography database was developed for these key basins and three adjacent basins (S-133, S-135, and S-154C). These seven study basins make up the primary drainage network for area stormwater discharges to Lake Okeechobee.

Location map for the Lake Okeechobee watershed detailing individual drainage basins and the seven study basins.

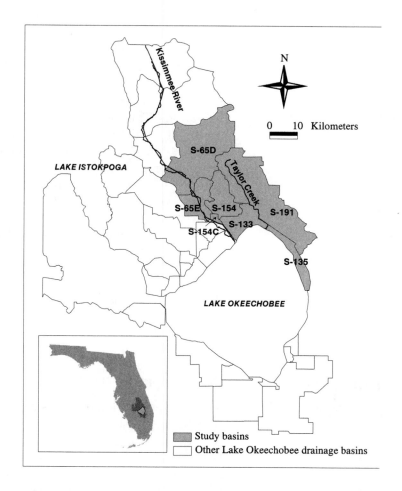

The parameters used to describe specific basins and respective stream characteristics are land use, stream flow conditions, vegetation type, and wetlands. The predominant land uses in the S-133 basin are urban, residential, and commercial. The predominant land use in the S-135 basin is agriculture with some rural development along the lake. The primary land use in the S-154, S-191, S-65D, and S-65E basins (the four key basins) is agriculture, a mix of dairy, beef, citrus, and row crop operations.

The major drainage in these basins is through large interceptor canals built to transport stormwater from secondary and tertiary systems in the Lake Okeechobee watershed. Flow rates are controlled by gated structures and pumps, and are therefore a function of gate openings and pump operations. Thin brush and Bahiagrass characterize the vegetation along the major canal banks. Secondary and tertiary drainage are conveyed by natural sloughs and creeks and manufactured drainage ditches. The manufactured tertiary drainage ditches feed the mostly natural secondary drainage systems. Some sloughs and creeks have been excavated to improve drainage conditions for agricultural and rural development. Stream flow within these systems ranges from no flow or dry conditions to flood stages that extend beyond stream banks into adjacent lands. The vegetation in these secondary and tertiary systems varies from thin brush to dense weeds. Thick stands of timber and underbrush characterize the hammock type sloughs while wetland plants dominate the marsh type sloughs.

Database Development

Hydrography Coverage

A GIS map containing hydrography, U.S. Geological Survey (USGS) quadrant names and boundaries, roads, and basin boundaries was created from the SFWMD's GIS database. Using this map, hydrography excluding field ditches was identified as either major, secondary, or tertiary drainage systems. This map was verified and descriptive attributes were added. Verification of hydrography information and the addition of vegetation related attributes in each basin were accomplished through a three-person review of numerous maps and subsequent consensus based on local field knowledge. Maps used in this review process included 1984 Mark Hurd aerial photographs, 1990 County aerial photographs, Okeechobee County Department of Transportation maps, and USGS 7.5-minute series topographic maps. Differences in interpretation among panel members were rectified through field and aerial inspections. Assigning name attributes to major, secondary, and tertiary drainage systems was achieved using the abovementioned sources along with information from the Okeechobee County plat records and the Okeechobee County Surface Water Management Master Plan. In areas where stream reaches were unnamed, current landowner names were assigned.

This GIS map served as a reference tool for the development of the hydrography database. Utilizing ARC/INFO, three coverages containing major, secondary, or tertiary hydrography were created from this map. Nodes indicating the beginning and end of each stream reach were added to the coverages. The INFO tables of these coverages were altered such that each arc in a coverage could be identified as belonging to either a major, secondary, or tertiary system. The resulting coverages were then merged into a single coverage for further processing.

Attribute data containing stream name, vegetation type, roughness coefficient, and basin location for each reach in the merged coverage were created. The data were converted into a tabular ASCII file for editing. The ASCII file was then imported into a temporary INFO file and joined to the coverage's arc attribute table by a unique reach identifier.

Linkage with Cross-section Data

An important function of the GIS hydrography database is to supply the P transport model with cross-section data. Stream cross-section data for selected water quality monitoring sites were stored as AutoCAD drawings, generated from a site survey. An ARC/INFO point coverage was created to represent the location of each cross section on the stream network. The coverage's INFO file was modified to include information about the cross section location in a corresponding basin, the station name, and the cross-section number.

The ArcView hot link feature was used to display the cross sections at water quality monitoring sites. To use this feature, AutoCAD cross-sectional drawings were converted into Arc coverages, and each cross-section coverage was brought into ArcView as a theme. The hot link between cross-section themes and corresponding locations in the water quality station coverage required an additional column in the water quality station attribute table. This column contained the names of the cross-section themes and was cross referenced with the coverages' cross-section numbers. The themes were then hot linked to the appropriate point in the water quality station coverage. Using the ArcView hot link feature, a user can point and click on a monitoring site and the cross section at that site will be opened for display.

Major, secondary, and tertiary streams.

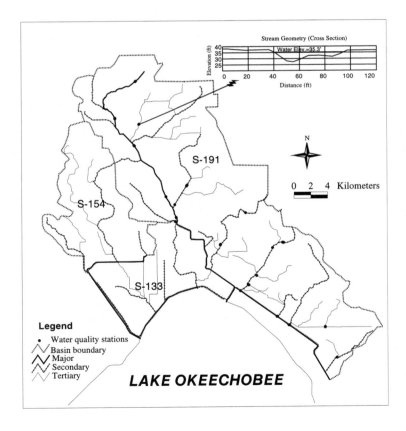

Stream Reach Numbering

Another important function of the GIS hydrography database is to supply the P transport model with stream network information. The stream can be conceptualized as a string of reaches that are linked sequentially to one another with pre-assigned numbers. The model requires the reach numbers to be unique, continuous, and sequential from upstream to downstream. When a stream branches, the reach numbering sequence shall reflect the stream level (major, secondary, or tertiary) of the branching sections. At the confluence, the higher level upstream reach is numbered first, and the lower level tributary reach is numbered last. The next figure illustrates the reach numbering scheme with a simplified stream network.

The Lake Okeechobee watershed includes 33 sub-basins, each with a stream network containing many reaches. Assigning reach numbers to these stream networks is a tedious and time-consuming task, if performed manually. Whenever the hydrography coverage is updated, the reaches must be renumbered to reflect the changes. To address this issue, a hydrographic topology analysis program (Hydro-Topo) was developed using AML, ARC/INFO's macro language. Streams are represented as lines (arcs) in an ARC/INFO coverage. The branching points and locations where hydrogeometric properties change along a stream are depicted as nodes in the coverage. A stream starts at a node, ends at a node, and is separated into multiple reaches by nodes. A stream reach is defined as a section with a unique combination of geometry and vegetation. The program detects the arc-node topological relationship in a stream network, checks the arc attribute table for information on network topology and vegetation type, and automatically assigns reach numbers for the entire network. The program limits the stream network to a single outlet. If a network contains multiple outlets, it is broken into subnetworks, each containing a single outlet.

*Simplified stream
network
illustrating
the reach
numbering
scheme.*

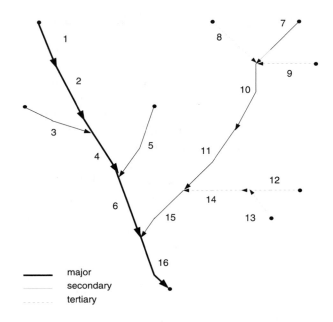

Depicted in the next illustration is the decision making process for the Hydro-Topo program. The program examines all
nodes in a stream network to identity the outlet node.
Because this node has no arc, it is left in a downstream direction. Starting from the outlet node, the program parses
upstream through the network, one arc at a time to assign
reach numbers. All arcs ending at a node and lacking an
assigned reach number are identified. If only one arc is identified, the program assigns a descending sequential reach
number to the arc, and then picks up the beginning node of
that arc to move further up the stream network. If more than
one arc is selected, arcs are sorted by stream level as the primary key and an ARC/INFO assigned record number as the
secondary key. The first arc in the sorted list is processed as
if it were the only arc ending at the node. If all arcs upstream
of a node are assigned reach numbers, the program searches
for a downstream arc. If an arc is found, the program moves
to the arc's end node and on to other branches. If the node
has no downstream arc, the search returns to the stream outlet node, and the program terminates.

Processing flow diagram for hydrographic topology analysis.

The Hydro-Topo program also helps database managers to check the quality of the hydrography coverage. A coverage will fail the quality examination if any of the following conditions are found to be true: (1) an arc is not connected to the network; (2) a pseudo node separates a stream section into two reaches with an identical attribute and without one or more branches; (3) two connected arcs point toward each other, or point away from each other; (4) the arc or node topology is not properly built; and (5) an arc does not have the required attributes (network topology and vegetation type).

Results and Discussions

Results of this study are best summarized in the form of GIS plots. Depicted in the second illustration above is a three-level hydrography (major, secondary, and tertiary) for three basins in the Lake Okeechobee watershed. The hydrology and water quality for these streams can be simulated using the P transport model. As a result, streams with higher P loading rates and fields that contribute high P loads to these streams can be identified. Further investigation would then be needed to evaluate the effectiveness of the best management practices in reducing P loads in these fields and streams.

Stream cross-section data can be viewed at the selected water quality monitoring sites (see the same illustration). These data are stored in a format that is directly usable by HEC-2. In turn, HEC-2 provides the hydraulic data for the P transport model.

The hydrography data are the basis for defining reach numbers for the P transport model. In P transport modeling, the stream can be conceptualized as a string of reaches that are linked sequentially to one another via the mechanisms of transport and dispersion. A reach has the same hydrogeometric properties: slope, cross section, roughness coefficient, and so forth. Using the Hydro-Topo program, assigning reach numbers to a network becomes a simple task.

The attribute table of the hydrography coverage contains parameters (stream names, reach length, and roughness coefficients) used by the P transport model (see following table). The reach length x can be used to derive the reach's computational elements, each of length Δx. The hydrologic and nutrient balances were conducted for each computational element in P transport modeling. Value ranges of roughness coefficients were obtained from Jobson and Froehlich (1986).

Example of hydrography coverage attribute data (model parameter values for the P transport model)

Stream name	Length (km)	Vegetation code*	Roughness coefficient
Taylor Creek	8.2	Ce1	0.05-0.12
Taylor Creek	2.5	Ce1	0.05-0.12
Taylor Creek	2.4	Ce1	0.05-0.12
Taylor Creek	0.4	Ce2	0.04-0.08
Taylor Creek	2.3	Ce2	0.04-0.08
Taylor Creek	7.2	Ce2	0.04-0.08
Taylor Creek	2.0	Ce2	0.04-0.08
Rim Canal	2.4	Cc2	0.035-0.06
Rim Canal	2.3	Cc2	0.035-0.06
Rim Canal	3.7	Cc2	0.035-0.06

Vegetation codes were defined based on channel and vegetation types. Ce1 represents an unmaintained manufactured ditch with dense weeds as high as flow depth. Ce2 represents an unmaintained manufactured ditch with clean bottom and brush on sides. Cc2 symbolizes a manufactured, very deep canal with thin brush at sides.

The GIS hydrography database will be linked with the P transport model using AML programs to create a seamless environment for data preparation, model execution, and output display and analysis. The phosphorus transport model used in conjunction with a GIS hydrography database provides a tool for researchers, planners, and regulators to evaluate water quality improvements as a result of agricultural best management practices. Researchers continue to improve model output and expand model applications as more data are generated, evaluated, and incorporated into the model. Planners can use output scenarios to identify the spatial extent of high phosphorus source areas within the Lake Okeechobee basin and develop strategies to minimize or eliminate these source areas. In addition, the regulators can use the model to evaluate the effectiveness of site specific management strategies for reducing P loads from permitted land parcels.

Summary

To prepare the hydraulic data efficiently for the P transport model, a GIS hydrography database was developed. This database contains three types of data used by the P transport model: (1) a GIS coverage including the hydrography and associated attributes for major, secondary, and tertiary streams; (2) stream names, geometry (cross-section data), vegetation types, and roughness coefficients; and (3) stream reach numbers derived from a GIS algorithm based on stream network topology and associated attributes.

In this GIS database package, the user can point and click on a cross-section location and the cross section will be displayed using the ArcView hot link feature. An AML (ARC/INFO's macro language) program, Hydro-Topo, for stream network topology analysis was also developed. This program detects the arc-node topological relationship in a stream network, checks the arc attribute table for information on network topology and vegetation type, and automatically assigns reach numbers for the entire network. Assigning reach numbers is an important step for input data preparation in P transport modeling. This GIS hydrography database will be linked with the P transport model to provide a tool for users to evaluate water quality improvements as a result of agricultural best management practices.

Acknowledgments

The authors would like to thank Todd Tisdale, Alan Steinman, Kurt Saari, Sue Hohner, and Barry Rosen for their valuable comments on an earlier version of this essay.

References

Brown, L.C. and T.O. Barnwell. "The enhanced stream water quality model QUAL2E and QUAL2E-UNCAS: documentation and user's manual." EPA/600/3-87/039. Athens, Georgia: U.S. Environmental Protection Agency, 1987.

Heatwole, C.D., K.L. Campbell, and A.B. Bottcher. "Modified CREAMS hydrology model for coastal plain flatwoods." *Transactions of the ASAE* 30:4 (1987), 1014-22.

_____. "Modified CREAMS nutrient model for coastal plain flatwoods." *Transactions of the ASAE* 31:1 (1988), 154-60.

Jobson, H.E. and D.C. Froehlich. "Basic hydraulic principles in open channels." NSTL, Mississippi: Gulf Coast Hydroscience Center, U.S. Geological Survey, 1986.

South Florida Water Management District (SFWMD). "Surface water improvement and management (SWIM) plan update for Lake Okeechobee." West Palm Beach, Florida: SFWMD, 1993.

_____. "Surface water improvement and management (SWIM) plan update for Lake Okeechobee." West Palm Beach, Florida: South Florida Water Management District, 1997.

U.S. Army Corps of Engineers. "HEC-2 Water Surface Profiles User's Manual." Davis, California: Hydrologic Engineering Center, 1990.

Wagner, R.A., T.S. Tisdale, and J. Zhang. "A framework for phosphorus transport modeling in the Lake Okeechobee watershed." *Water Resources Bulletin* 32:1 (1996), 1-17.

Zhang, J., T.S. Tisdale, and R.A. Wagner. "A basin scale phosphorus transport model for south Florida." *Applied Engineering in Agriculture* 12:3 (1996), 321-27.

Indicators of Resources and Landscapes

*R.J. O'Connor, University of Maine, and P.R.H. Neville
and T.B. Bennett, University of New Mexico*

This book has introduced ideas for compiling data pertinent to a natural resources GIS, as well as where to obtain data and how to build databases. However, GIS layers developed independently may not fully produce solutions to the problems posed by an everchanging environment, nor address complications imposed by changes in cultural, socioeconomic, and regulatory conditions. A successful GIS is adaptable to various conditions, maximizing the potential of newly derived information using a combination of existing data sets and dynamic variables. This chapter will touch on some of the tools available to derive this information in order to provide the framework necessary for developing solutions to natural resource management questions.

Given a remotely sensed image or GIS-derived representation of appropriately processed landscape components, the various attributes of the scene–land cover, soil types, vegetation state, roads, rivers, and so on–can be combined to yield resource or landscape indicators. Indicators can be used in two roles: (1) to index the condition of a landscape, and (2) to detect changes over time of interest to resource managers.

An example of using indicators in these roles is quantifying an area for its suitability as wildlife habitat. This is typically accomplished by filtering and combining key features of the scene that are correlated with the known niche requirements of the species. The development of a spotted owl habitat model for the Pacific Northwest (Gutierrez and Carey 1985) has become a classic of this type of approach. Once this habitat index has been developed for different times, the changes in wildlife habitat availability can be tracked, thereby mapping such factors as forest, grassland succession, or changes in land use. Given the relative youth of satellite remote sensing, change detection of natural resources across years is as yet little used; Muchoney and Haack (1994) and Chavez and MacKinnon (1994) provide some examples. Change detection using aerial photography, on the other hand, has a longer history.

Several broad classes of indicators deserve discussion. These indicators can be used to characterize *composition*, *pattern*, *shape*, and *scale*. Composition refers to the proportional presence of land cover classes, such as forest, cropland, prairie, wetland, and water, among others. Pattern is the spatial distribution of class composition and is important because landscapes of equal size and composition can provide different habitat structure for organisms, depending on varying surroundings and shapes. The determination of effective scale is also important for creating accurate models of the natural resource of interest. For instance, a mosaic (e.g., wooded savannah) has a unitary identity and scale quite different from those of its constituent woodland and grassland patches. In brief, a mosaic model scales differently from a woodland/grassland patch model.

Composition Metrics

Composition is expressed as the proportion (percentage) of a spatial unit covered by a particular land cover class. A variety of land cover classification schemes exist, differing mostly in levels of detail. At a coarse level, the Anderson Level I classification (Anderson et al. 1976) provides nine land cover classes (urban or built-up, agricultural, rangeland, forest, water, wetland, barren, tundra, perennial snow or ice). In this scheme forests of all types are lumped into a single class. In contrast, the prototype land cover classification developed by Loveland et al. (1991), using green-up profiles derived from spectral data, contains 159 classes, of which 59 are forest covers of various types. The choice of a classification strategy is driven by the particular problem and the intended use of the resulting indicators.

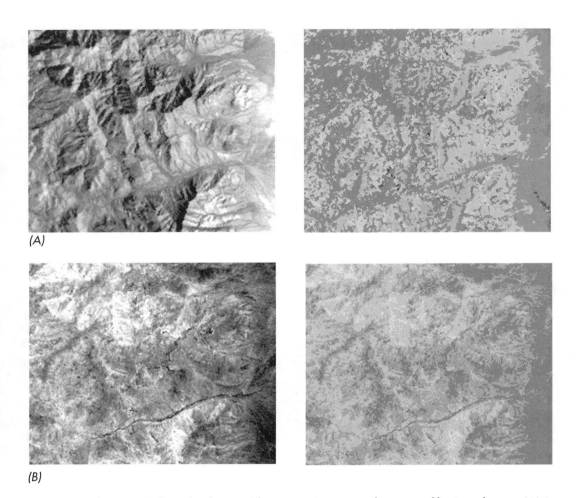

(A)

(B)

Compositional metrics. A desert landscape with a mountain range to the west and basin to the east. (A) At the left, grayscale version of a natural color display of a Landsat Thematic Mapper image. At the right, a supervised classification of the image: woodland classes in darker grays (mostly at left), grasslands in whitish patches (middle right), and shrublands in medium and dark grays on right side and as islands within grasslands. (B) Vegetation index on the left with the brighter values representing high biomass. On the right, the same image has been density sliced to represent four main groups ranging from black (moderately vegetated) through diagonal hatching to open stipple (barren).

Many existing data sets are classified and ready for direct implementation into a natural resources GIS. These range from polygons attributed with ownership status through stream networks indicating stream order to point locations of archeological sites differentiated by historical period. Other data sets may require reaggregation in order to be useful. Examples include a percentage slope coverage recoded to degrees of erodibility or a vegetation index recoded to percent cover (see next illustration). Using multivariate data sources such as satellite multispectral data, a classification can be created using supervised or unsupervised clustering methods with a classification decision rule such as maximum likelihood.

Pattern Metrics

Pattern metrics are intrinsically more complex than composition metrics. Landscapes of any interest typically contain three structural or pattern element: *patches, corridors,* and an underlying *matrix* (see next illustration). Patches or *ecotopes* are distinct spatial or geographic areas that can be distinguished from their surroundings, which may be either a background matrix of a single cover class or a mosaic of contiguous patches of diverse cover classes. Thus, one may see water bodies set in a matrix of wheat in an agricultural setting, or a checkerboard mosaic of patches of different emergent vegetation spread over a shallow wetland. Individual patches may vary in size and shape, a point of some importance for animals found in these habitats. Where individual patches of different type abut, an *ecotone is* typically associated with a local increase in wildlife diversity. Finally, individual patches may not be isolated from one another but may be linked by *corridors* which themselves may be regarded as elongated "patches" of habitat that link one patch to another. The full complexity of natural landscapes defies summary in a small number of pattern indicators. Some major groups of indices are listed in the following paragraphs.

(A)

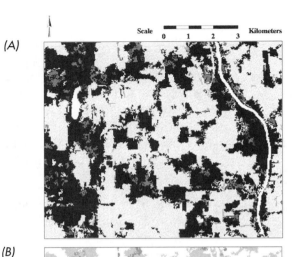

Structure in landscapes. (A) A video image of a landscape in Aroostook County, Maine, showing the interspersion of forest (black) and agricultural land (gray). (B) The same image retouched to show forest examples of patch and corridor set in a matrix (of agricultural land in this example). Note also how the river to the right fragments blocks of forest and farmland.

(B)

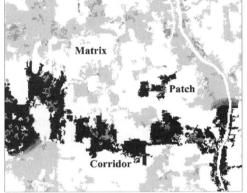

Dominance and Diversity Indices

Dominance specifies the proportion of the area covered by the largest single land cover, irrespective of its identity. Thus two spatial units may each have dominance values of 0.85, but in one case be 85% under corn fields and in the other be 85% under conifer plantations. Similarly, diversity indices provide complementary measures of dominance, indicating the diversity of patch types within a spatial unit of analysis. Clearly, landscapes with high dominance are likely to be permeable to species that can use the dominant habitat, and unfavorable to those that cannot, whilst mosaic habitats favor only species that can generalize across habitats to view the mosaic as a connected network. Specialized filters can be used to quantify the patterns for a thematic map or classification (see next illustration). When dealing with non-thematic data such as a satellite image or a vegetation index, a statistical tool such as a variance or standard deviation filter can be an effective diversity indicator. Various contagion metrics quantify the extent to which land covers coalesce to form larger patches, although the diversity of such metrics, often called "contagion," requires care when resorting to a formula.

(A) *(B)* *(C)*

Dominance and diversity indices. (A) A dominance filtered image derived from the classification in the previous illustration. The brightest values represent the classes with the greatest number of pixels. (B) A diversity filtered image derived from the same classification. The brightest values represent areas with the most heterogeneous land cover. (C) A variance image of the vegetation index in the first illustration. The brightest values represent areas with the greatest change in overall vegetation cover.

Connectivity Indices

Connectivity of landscapes is important when evaluating movement phenomena, such as when organisms traverse a landscape, fire spreads across a forest, or flooding spreads across terrain. One very general finding is that most randomly organized landscapes have a critical connectivity value for habitat at around 59% (O'Neill et al. 1992). In other words, once the proportion of the landscape covered by the habitat of interest exceeds this threshold it is virtually certain that an organism can get from one side to the other of the landscape without ever leaving the specified habitat. Particular configurations of habitat that violate this generalization are conceivable, but it is normally safe to conclude that this level of habitat ensures effective connectivity. Indexing particular landscapes according to whether natural or human dominated land cover exceeds the 59% threshold can then be considered. The former probably still retain a high degree of natural wildlife communities (O'Neill et al. 1997). Conversely, the relative connectivity of the landscape also indexes its vulnerability to the spread of fire, disease, and flooding. High connectivity habitats are more prone to very large-scale incidents (Jones et al. 1996).

Schumaker (1996) has conducted the only systematic assessment of how well connectivity indices predict dispersal success. He found that all nine of the commonly used connectivity metrics behaved poorly in this regard but that a new measure he called "patch cohesion" was superior. Patch cohesion is a standardized measure of a quantity he termed ξ, which is the area-weighted mean value of the perimeter-area ratios in a spatial unit divided by the area-weighted mean of the corresponding shape index. Patch cohesion for raster data follows:

$$PC = [\xi(max) - \xi]/[\xi(max) - \xi(min)]$$
$$= [1 - \Sigma p/\Sigma(p\sqrt{a})]/[1 - 1\sqrt{N}]$$

where $\xi(min)$ is the minimum of ξ, obtained when one patch fills the landscape in the grid unit, and $\xi(max)$ is the maximum of ξ, obtained when all patches are single pixels. Variables $a = A/s^2$ and $p = P/s$ are the area and perimeter of a patch in pixels and pixel edges, respectively; A and P are the original area and perimeter, and N is the total of (all) pixels in the landscape (grid unit).

Fragmentation Indices

Habitat fragmentation is the process of dissecting large and contiguous areas of similar native vegetation types into smaller units separated by different land use, often by human dominated ones. Fragmentation involves direct loss of habitat and isolation of the remaining blocks of habitat as patches. In the early stages of fragmentation, patches are often still connected by corridors of habitat along which species can recolonize patches. The density of roads, and especially new roads, has been used as an index of fragmentation since roads are often the first indicator of habitat dissection. An example of this is the increase in road and housing densities over a 50-year period as an index of fragmentation of elk wintering habitat (W. Dunn personal communication). Intensified fragmentation removes corridors, and patches become smaller and more isolated. It becomes increasingly difficult for all but the most mobile species to recolonize vacant patches. Such change can be tracked by following the total amount of the habitat present and by following patch size distributions in a time series of landscape images. Comparing composition data and patch size distributions between regions can be useful indicators of the relative impacts of habitat loss and fragmentation. O'Neill et al. (1997) compared land cover composition of regions against Küchler's (1964) map of potential natural vegetation and found that the areas least impacted were primarily mountainous regions and still had land cover comparable to that of Küchler, whereas heavily impacted regions had very different land cover.

Shape Metrics

Because human dominated landscapes have disproportionate representations of regular shape (fields, streets, canals), indices of shapes can provide information on the extent of human influence. *Shape metrics* quantify the shape of patches and other landscape elements, either as area-perimeter ratios or by relating shapes to standardized shapes such as circles or rectangles. Some quite simple metrics have been successfully deployed as ecological indicators. Vogelmann (1995) used the natural logarithm of forest area to perimeter ratio, dubbed a *forest continuity index*, to assess pattern and trends in forest fragmentation over southern

New England. Many species, and particularly bird species, are unable to continue as breeders in such dissected habitats, largely because predators and parasites (e.g., cowbirds) can more easily penetrate small compared to large forests. Vogelmann found that forest continuity decreased sharply with increase in the human population of the area until densities of about 200 people per square kilometer were reached, stabilizing thereafter. The simple shape index used by Vogelmann—essentially an area-perimeter ratio—may have been especially suited to the type of fragmentation occurring in New England. Residential development and associated roads start to dissect a forest by inserting long, narrow, cleared areas that are especially amenable to indexing by an area-perimeter ratio. Whether the same index would work as well in areas of extensive clear-cutting in the western United States is another question.

More complex measures of shape may be needed to index special situations. Linear habitats such as beaches, barrier islands, and mountain ridgelines can provide a "stepping stone" facility for wildlife to span the full range of habitat. In such situations, the loss of a single patch (beach, island, or ridgeline) can break the effective continuity of the chain and limit a species' use of the area. Similarly, the relative extent to which cowbirds can penetrate a patch of forest may be limited by the specifics of the patch's shape, if the cowbird will penetrate only a given distance from the forest edge. Computer simulation, either through direct modeling or by use of cellular automata, has been used to characterize such landscapes as to suitability for wildlife. The cellular automaton approach is particularly versatile: model "organisms" are released onto the landscape of interest and are allowed to move freely across pixels of ideal habitat and less freely across inferior or unsuitable habitat, and the cumulated results from the random release of many organisms provides a picture of the effective connectivity of the landscape. However, this technique is computationally costly.

Fractal dimension has been used to characterize land cover texture and the regularity or irregularity of patch perimeters. However, one should note that fractal dimension is mathematically related to semivariance and is therefore (or perhaps truly is) an index of spatial autocorrelation. Fractal dimension is calculated as regressions of perimeter to area ratios, revealing its relationship to simpler shape complexity measures.

Scale Metrics

To the human eye, a forest stand of a particular species might constitute a single patch. To a moth, particular clusters of (or even single) trees within the forest might be recognizable as potentially poor or potentially very good laying areas. To the resulting caterpillar, different branches or even individual leaves might be readily discriminated as patches of different quality. New technologies have not necessarily overcome this bias as different sensor technologies will see the same scene differently and what is significant to one sensor may be unresolvable to another (see next illustration). The scale problem has relevance beyond species studies: biodiversity is often equated with the number of species in an area, but in reality extends beyond simple numbers to encompass genetics and ecosystem diversity. Hierarchy theory and several empirical studies (Holling 1992) suggest that species and their interactions should occur at discrete scales of size and space. Hence, measuring the number of scales present in a landscape may be indicative of its biodiversity quality. Scales can be determined statistically or as the occurrence of distinct faunal groups or even biological provinces (Jones and Riitters 1996, Ray et al. 1998).

How scale affects perception in forested agriculture in Aroostook County, Maine. (A) Zoomed video image of a forested headland. (B) Same area (inset) in context of a wide-angle video shot. (C) Wide-angle image (inset) within a 1991 Thematic Mapper image derived from Band 4 (dark color for forest, gray color for agriculture).

(A)

(B)

(C)

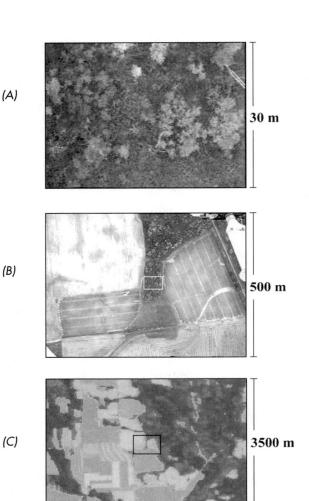

30 m

500 m

3500 m

The issue of scale is not only a spatial phenomenon since time, and how much temporal variability should be considered for the natural resources model, are also important considerations. Pickett and White (1985) showed that there is direct correlation between size and frequency of disturbance. Most studies of an area occur within a couple of years and may not adequately integrate the effects of large, albeit infrequent, disturbance events. In the southwestern

United States, many management decisions are based on well-understood annual variations in precipitation and temperature but fewer decisions incorporate the effects of periodic El Niño/La Niña events. Other climatic cycles which occur on the scale of decades can have even farther reaching effects. Indeed, some biotic systems are developed and maintained by a certain amount of disturbance, so that models incorporating periodic fires, floods, or other events (e.g., insect outbreaks) in ecosystems would better represent natural functions.

Modeling time effects can be especially difficult when looking at dynamic processes such as climate or ecosystems. Statistical tools such as principal component analysis (Eastman and Fulk 1993) or temporal indices (Samson 1993) are among the many options for modeling these effects. Additional possibilities exist in using the patch size distribution across different land classes as an indicator of the optimal time cycle for each successive stage of an ecosystem in equilibrium.

Discussion

Many of these metrics, when calculated for a common data set, turn out to be mutually correlated. Riitters et al. (1995) computed 58 metrics for U.S. Geological Survey Land Use Data Analysis (LUDA) maps, but only 26 of these were mathematically independent, and only six statistically independent axes of variation existed within the data set. These six axes of variation together accounted for 87% of the variation among the 26 indicators, and respectively describe a composite measure of patch compaction, overall texture, average patch shape, patch perimeter-to-area scaling, number of classes, and scaling of large patch density to area.

Cain et al. (in press) tested metric independence by using high resolution (25m) Thematic Mapper (TM) data for the Tennessee River and Chesapeake Bay watersheds, and found that six factors derived from 28 variables captured most of the variation. Three factors indicated texture, patch perimeter-to-area scaling, and number of cover types to be independent axes of variation, but measures of patch shape and patch compaction varied in odd ways, apparently as a result of changes in scaling undertaken within the study. Perry and Lautenschlager (1984), in their synopsis of vegetation indices used with satellite data, found that many of the indices are highly correlated with each other. These studies show that many land cover indicators are redundant even across multiple scales, and few truly independent dimensions of variation in landscapes exist. Therefore, considering the time and resources that could be spent on generating these indicators, it is advisable that the natural resource manager consider what precisely is needed for the problem at hand and be parsimonious with the indicators used.

Conclusion

Turner and Gardner (1991) and O'Neill et al. (1988) review the pattern and scaling metrics available for landscape characterization and provide good reviews of technical issues associated with different approaches. The formulas for many of the landscape indicators developed to characterize landscapes on the basis of mapped data are listed in Riitters et al. (1985) and O'Neill et al. (1988). Some of the most important environmental degradation in our time occurs at landscape scales, with intensification of forestry, river channelization, and agriculture leading to clear-cutting of forests, loss of wetlands, and extensification of cropland and rangeland. Because many of these processes occur in a spatial context, the use of these metrics holds major promise in tracking and understanding these disruptions.

Acknowledgment

We thank Randy Boone (University of Maine) for providing the second and fourth illustrations.

References

Anderson, J.R., E.E. Hardy, J.T. Roach, and R.E. Witmer. "A land use and land cover classification system for use with remote sensor data." Geological Survey Professional Paper 964. Washington, D.C.: U.S. Geological Survey, 1976.

Cain, D.H., K. Riitters, and K. Orvis. "A multi-scale analysis of landscape statistics." In *Landscape Ecology,* forthcoming.

Chavez, P.S., and D.J. MacKinnon. "Automatic detection of vegetation changes in the southwestern United States using remotely sensed images." *Photogrammetric Engineering & Remote Sensing* 60:5 (1994), 571-83.

Eastman, J.R. and M. Fulk. "Long sequence time series evaluation using standardized principal components." *Photogrammetric Engineering & Remote Sensing* 59:6 (1993), 991-96.

Gutierrez, R.J. "An overview of recent research on the spotted owl." In R.J. Gutierrez and A.B. Carey, eds., *Ecology and Management of the Spotted Owl in the Pacific Northwest,* pp. 39-49. Washington, D.C.: U.S. Department of Agriculture Forest Service General Technical Report PNW-185, 1985.

Holling, C.S. "Cross-scale morphology, geometry, and dynamics of ecosystems." *Ecological Monographs* 62 (1992), 447-502.

Jones, B., J. Walker, K.H. Riitters, J.D. Wickham, and C. Nicholl. "Indicators of landscape integrity." In J. Walker and D.J. Reuter, eds., *Indicators of Catchment Health: a Technical Perspective,* pp. 155-68. Melbourne: CSIRO, 1996.

Jones, K.B., and K.H. Riitters. "Evaluating wildlife habitat suitability using a multi-scaled landscape assessment approach." *Computing Science and Statistics* 27 (1996), 140-46.

Küchler, A.W. "Manual to accompany the map: potential natural vegetation of the conterminous United States." New York: American Geographical Society, Special Publication 36, 1964.

Loveland, T.R., J.W. Merchant, D.O. Ohlen, and J.F. Brown. "Development of a land-cover characteristics database for the conterminous U.S." *Photogrammetric Engineering & Remote Sensing* 57 (1991), 1453-63.

Muchoney, D.M. and B.N. Haack. "Change detection for monitoring forest defoliation." *Photogrammetric Engineering & Remote Sensing* 60:10 (1994), 1243-51.

O'Neill, R.V., R.H. Gardner, M.G. Turner, and W.H. Romme. "Epidemiology theory and disturbance spread on landscapes." *Landscape Ecology* 7 (1992), 19-26.

O'Neill, R.V., C.T. Hunsaker, K.B. Jones, K.H. Riitters, J.D. Wickham, P. Schwarz, I.A. Goodman, B. Jackson, and W.S. Baillargeon. "Monitoring environmental quality at the landscape scale." *Bioscience* 47 (1997), 513-20.

O'Neill, R.V., J.R. Krummel, R.H. Gardner, G. Sugihara, B. Jackson, D.L. DeAngelis, B.T. Milne, M.G. Turner, B. Zygmunt, S.W. Christensen, V.H. Dale, and R.L. Graham. "Indices of landscape pattern." *Landscape Ecology* 1 (1988), 153-62.

Perry, C.R., Jr. and L.F. Lautenschlager. "Functional equivalence of spectral vegetation indices." *Remote Sensing of Environment* 14 (1984), 169-82.

Pickett, S.T.A., and P.S. White. *Patch Dynamics: A Synthesis, The Ecology of Natural Disturbance and Patch Dynamics*. New York: Academic Press, 1985.

Ray, G.C., B.P. Hayden, M.G. McCormick-Ray, and T.M. Smith. "Land-seascape diversity of the U.S. east coastal zone with particular reference to estuaries." In R.F.G. Ormond, J.D. Gage, and M.V. Angel, eds., *Marine biodiversity: patterns and processes*. Cambridge: Cambridge University Press, 1998.

Riitters, K.H., R.V. O'Neill, C.T. Hunsaker, J.D. Wickham, D.H. Yankee, S.P. Timmins, K.B. Jones, and B.L. Jackson. "A factor analysis of landscape pattern and structure metrics." *Landscape Ecology* 10 (1995), 23-39.

Samson, S.A. "Two indices to characterize temporal patterns in the spectral response of vegetation." *Photogrammetric Engineering & Remote Sensing* 59:4 (1993), 511-17.

Schumaker, A.C. "Using landscape indices to predict habitat connectivity." *Ecology* 77 (1996), 1210-40.

Turner, M.G. and R.H. Gardner. *Quantitative Methods in Landscape Ecology: The Analysis and Interpretation of Landscape Heterogeneity.* New York: Springer-Verlag, 1991.

Vogelmann, J.E. "Assessment of forest fragmentation in southern New England using remote sensing and geographic information systems technology." *Conservation Biology* 9 (1995), 439-49.

Global Vegetation Production and Human Activity

S.D. Prince and M.E. Geores, University of Maryland

Resource Management Requirement

Earth orbiting satellites have furnished a wealth of data on natural resources at various spatial and temporal resolutions. These data are now able to measure global vegetation primary production, a particularly sensitive variable that indicates the role of land cover for fixing carbon (Goward and Prince 1995). Because fixed carbon constitutes the source of food, fuel, and fiber for much of the world's population, it has fundamental significance for understanding the human dimensions of natural resource supply and demand. Not only does primary production indicate crop, forest, and rangeland output, but it is also a sensitive indicator of land surface condition and can be used to detect degradation (Prince et al. 1998). This case study is part of an ongoing investigation of the relationship between human dimensions of natural resource availability and primary production on a global scale.

Numerous recent studies have modeled the Earth's potential primary production in the absence of human activities. These models vary from highly parameterized, detailed models to statistical models forced with simple climatological observations. For a discussion of the former, see Woodward et al. (1995), Neilson (1995), and Haxeltine and Prentice (1996). The latter are addressed by Lieth (1975). One simple model reported by Goward and Prince (1995) maximizes the advantages of global satellite observations to parameterize a mechanistic climate model of global production. The model was used here in conjunction with satellite data to determine the difference between actual and potential primary production using a global 0.5° x 0.5° grid.

The results of the global comparison are intriguing. They suggest that parts of the land surface are functioning at close to their climatically determined maximum, while production in others is a fraction of potential. Explaining these patterns requires a combined human and biophysical approach. Not only are there many potential explanations but the analysis also must be conducted at a spatial scale appropriate to a 0.5° x 0.5° grid without misusing the information in any of the human, physical, or biological domains involved.

Methodology

Current satellite data sets produced from the National Oceanic and Atmospheric Administration's Advanced Very High Resolution Radiometer (AVHRR) sensor permit studies at temporal scales from about 10 days to 16 years and spatial scales from 20 sqkm to those encompassing the entire globe (Townshend 1992). Many major patterns of vegetation activity and structure can be observed in AVHRR sensor data. Details of these observations are reported in Townshend and Tucker (1984), Tucker et al. (1985), Malingreau (1986), and Goward et al. (1987). In addition, vegetation cycles, including interannual variations in net primary production (NPP), may be derived from such observations. The patterns are correlated with climatic variables (Prince 1991, Prince and Goward 1995). Although daily observations are collected, cloud cover obscures a significant percentage of all land areas on any given day. Typically, 10 days or more of daily observations are combined to produce nearly cloud-free composite views of land areas.

The remote sensing measurement most commonly employed in vegetation studies is a visible/near infrared spectral vegetation index (SVI). The marked contrast between strong absorptance in visible wavelengths and strong reflectance in near infrared wavelengths uniquely characterizes the presence of pigmented, photosynthetically active foliage in terrestrial landscapes. There are many alternate SVI formulations but one of the most commonly used with the AVHRR observations is the normalized difference vegetation index (NDVI).

SVI measurements are related to leaf area index, green biomass, percentage of ground cover, and fraction of incident PAR absorbed in canopies for various vegetation types. With the possible exception of the latter variable, the observed relationships vary across vegetation types. In addition, there are several factors in the satellite remote sensing process that can disrupt precise interpretation of spatiotemporal patterns in vegetation systems on regional and global scales (Goward and Dye 1996). Nevertheless, analyses of the AVHRR satellite data show strong correspondence to vegetation. In particular, the summed area below the NDVI-time curve (referred to as ΣNDVI.days) is strongly related to primary vegetation production (Prince 1991). Correlations of ΣNDVI.days with mean annual temperature or total annual precipitation produce a structured but poorly correlated distribution. This is not surprising because temperature and precipitation can limit vegetation growth.

However, a simple "limiting factors" approach can be used to relate ΣNDVI.days to temperature and precipitation. For each climate variable, a bounding line can be defined joining the maximum ΣNDVI.days values for any value of the climate variable. These two relations can then be used to define an anticipated temperature-ΣNDVI.days and precipitation-ΣNDVI.days for each terrestrial grid cell. By selecting the lower of the two predicted values (i.e., the limiting factor), a simple, strong, linear correlation is found between the predicted ΣNDVI.days and the observed ΣNDVI.days (Prince 1991). The predicted ΣNDVIdays, derived from climatological data, represents potential vegetation production limited only by biophysical factors, while the observed ΣNDVI.days represents actual primary production. The difference between the two values can be expected, at least partly, to represent the impact of humans on global vegetation.

Results

The actual production map (item A in the following illustration) is based on a calculation of annual ΣNDVI.days from the satellite observations of the land surface for each 0.5° x 0.5° grid cell. The map for potential production (item B in the figure) shows the annual ΣNDVI.days modeled from climatological temperature and precipitation data. The difference, item C in the figure, contrasts potential with actual production and highlights regions where production was above the level predicted by the model (e.g., because of agricultural intensification techniques such as fertilization and irrigation), below the level (factors for which are explored below), or at the same level.

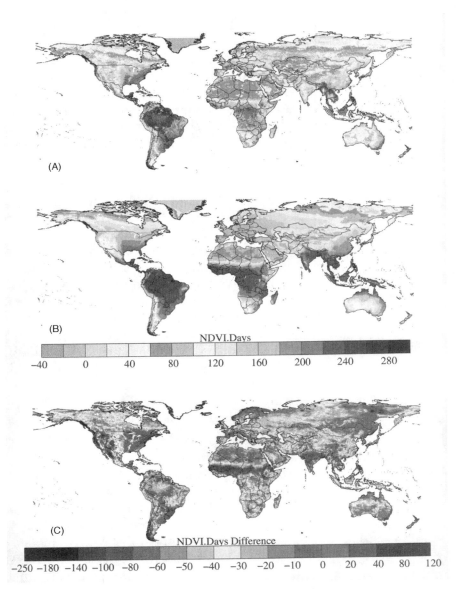

Global primary production: actual (A), potential (B), and difference between the two in ND-VI.days (C).

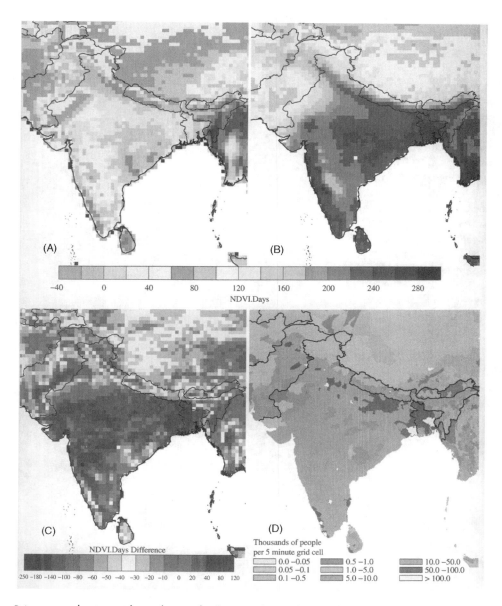

Primary production and population of India: population density of India (D), potential production (B), actual production (A), and the difference between actual and potential production in NDVI.days (C).

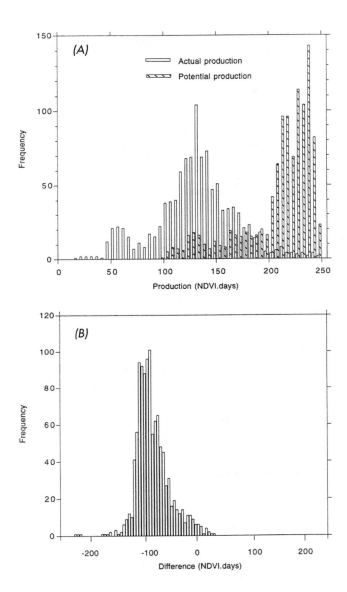

Frequency distribution of primary production for 0.5° grid cells in India. Actual versus potential (A), and the difference between actual and potential production in NDVI.days (B). The potential NDVI.days for India had a mean of 210, a median of 220, a range of 99 to 249, and a standard deviation of 34. The actual NDVI.days values were substantially lower, with a mean of 132, median of 133, range of 16 to 249, and standard deviation of 38.

Testing the hypothesis that human activity is the principal cause of deviations in production from the potential value in each grid cell presents a major challenge. The adopted starting point is the distribution of population, based on the proposition that land cover is altered most in those areas having the highest population density. Data for India provide an illustration of this hypothesis. Using the gridded population of the world developed by Tobler et al. (1995), the 1991 population of India was mapped by census district at a 5-minute grid resolution (see item D in the next illustration).

The image of potential production for India (item B in the next illustration) shows extremely high values in the northeastern area encompassing the Indo-Gangetic and west coastal plains, while the remainder of the subcontinent had lower values (but still high on a global scale). Actual production for India (item A) is far below its potential (item B); frequency distributions for potential, actual, and difference values (see the following figure) illustrate how markedly actual production deviates from the potential. The difference values are negatively skewed (s = -0.54), indicating an overall negative bias in actual and potential production in India.

Comparing human population density with production images, it is evident that areas with greater potential for production are more densely populated than other areas. However, it is also clear that the relationship between primary production and population is not constant—there are places in India, such as Hyderabad, where the difference between actual and potential production is less dramatic than in the Indo-Gangetic plain, despite high population density. This suggests that factors in addition to population should also be explored.

Other Factors

The global difference image (see above) confirms the necessity to consider additional descriptors of human activity. For example, if population density is the primary force driving down NDVI, then one would expect to find lower actual production in western Europe and the northeastern United States, where many people are concentrated. However, this is not the case. Conversely, much of Africa is not densely populated, but actual production falls well below potential.

Because population distribution alone cannot account for these differences, even in highly populated India, it was necessary to examine other descriptors of human activity.

Country level data are being used in this continuing study because of their availability, with the understanding that there may be regions within a country that differ significantly from the mean represented by this level of generalization. Comparing countries with high and low actual production suggests some interesting trends. The population structure, level of economic development, and intensity of agricultural practice are strongly related to actual production in regions having the same potential for production.

Production falls below what the model would indicate in countries where typically a large proportion of the population are under age 15. A high proportion of young people places a substantial drain on investment. Such countries experience higher infant mortality rates than countries where production exceeds model indicators, highlighting not only higher death rates but also poor access to technology (especially medical technology, the factor that most affects infant mortality rates). Somewhat surprisingly, countries where production falls far below potential have a less well-developed road infrastructure than do countries where production exceeds the potential. This suggests that access to an area by road is not a good indicator of land degradation. Countries with higher actual than potential production generally have higher incomes, measured in purchasing power parity/capita. These societies use substantially more fertilizer than do countries that do reach potential output levels. There are, of course, exceptions to these trends, making further analysis necessary.

Conclusions

Interpreting the impact of human beings on natural resources has previously concentrated on either finer resolutions (households) or coarser resolutions (countries) than the 0.5° x 0.5° scale considered here. Nevertheless, these observations suggest relationships between human activities and the difference between actual and potential vegetation production on a global scale. Thus, an inquiry at this unusually coarse resolution is justified, even though the variety of potential explanations and the impossibility of conducting controlled experiments make it difficult. Because the Earth's natural resources have become an issue of increasingly global concern, development of interpretations of phenomena observed at the global scale will increase in importance. International agreements on carbon emissions provide an immediate application for the findings.

Acknowledgments

We thank Charles Peters and Jonathan Haskett for assistance with data processing. This work was partly supported by NASA grants NAGW5178 to S.D. Prince and NAG53454 to S.N. Goward.

References

Esser, G. "Osnabrück Biosphere Model: Structure, Construction, Results." In *Modern Ecology: Basic and Applied Aspects,* edited by G. Esser and D. Overdieck, pp. 679-709. Amsterdam: Elsevier, 1991.

Goward, S.N. and D. Dye. "Global biospheric monitoring with remote sensing." In *The Use of Remote Sensing in Modeling Forest Productivity* edited by H.L. Gholtz, K. Nakane, and H. Shimoda, pp. 241-72. Dordrecht, The Netherlands: Kluwer Academic Publishers, 1996.

Goward, S.N. and S.D. Prince. "Transient effects of climate on vegetation dynamics: satellite observations." *Journal of Biogeography* 22 (1995), 549-63.

Goward, S.N., D.G.A. Kerber, and V. Kalb. "Comparison of North and South American biomes from AVHRR observations." *Geocarto* 2:1 (1987), 27-40.

Haxeltine, A. and I.C. Prentice. "BIOME3: an equilibrium terrestrial biosphere model based on ecophysiological constraints, resource availability, and competition among plant functional types." *Global Biogeochemical Cycles* 10 (1996), 693-709.

Malingreau, J.P. "Global vegetation dynamics: satellite observations over Asia." *International Journal of Remote Sensing* 7 (1986), 1121-46.

Neilson, R.P. "A model for predicting continental-scale vegetation distribution and water balance." *Ecological Applications* 5 (1995), 362-85.

Prince, S.D. "A model of regional primary production for use with coarse-resolution satellite data." *International Journal of Remote Sensing* 12 (1991), 1313-30.

Prince, S.D. and S.N. Goward. "Global primary production: a remote sensing approach." *Journal of Biogeography* 22 (1995), 815-35.

Prince, S.D., E. Brown de Colstoun and L. Kravitz. "Evidence from rain use efficiencies does not support extensive Sahelian desertification." *Global Change Biology* 4 (1998), 359-374.

Tobler, W., U. Deichman, J. Gottsegen, and K. Malloy. *The Global Demography Project, Technical Report 1995-6.* Santa Barbara, CA: National Center for Geographic Information and Analysis, 1995.

Townshend, J.R.G. and C.J. Tucker. "Objective assessment of AVHRR data for land cover mapping." *International Journal of Remote Sensing* 5 (1984), 492-501.

Townshend, J.R.G. *Improved global data for land applications: a proposal for a new high resolution data set.* Report of the land-cover working group of IGBP-DIS, No. 20. Stockholm: International Geosphere-Biosphere Programme, 1992.

Tucker, C.J., J.R.G. Townsend, and T.E. Goff. "African land-cover classification using satellite data." *Science* 227 (1985), 369-75.

Woodward, F.I., T.M. Smith, and W.R. Emanuel. "A global land primary productivity and phytogeography model." *Global Biogeochemical Cycles* 9 (1995), 471-90.

Using Population Data to Address the Human Dimensions of Environmental Change

D.M. Mageean and J.G. Bartlett, University of Maine

Resource Management Requirement

In recent years researchers and policy makers have identified population-environment interactions as crucial to issues of ecology, economic development, and human welfare. It seems clear that human populations and demands on the environment are driving ecological change in such areas as global warming, ozone depletion, deforestation, biodiversity loss, land degradation, and pollution of air and water. In developing countries, the anthropogenic effects include drastic environmental deterioration in areas where population pressures exceed a particular threshold (Terborgh 1989). In developed countries, environmental problems have arisen because increasing incomes, leisure, and ease of communication have generated a stronger demand for recreation and tourism (Bayfield 1979).

Enormous gaps still exist in the scientific understanding of precisely how demographic factors—such as population size and growth rate, settlement distribution, and migration dynamics—affect resources and the environment. For example, determining what proportion of environmental impacts result from population growth versus behavior (e.g., consumption patterns) is difficult. Furthermore, because of the fragmentation of research among disciplines, achieving a holistic picture is problematic.

In 1993, Paul Stern of the National Research Council described the need for a second environmental science that would be "focused on human-environment interactions—to complement the science of environmental processors by analyzing key questions" (Stern 1993). Stern outlined three main fields of inquiry for such questions: (1) the study of human causes of environmental change, (2) the effects of environmental change on things people value, and (3) the study of respective feedback between humanity and the environment. Unfortunately, little unanimity of opinion regarding the nature of the relationship hampers efforts in the first field of inquiry. As recently as 1992, Daniel Hogan wrote that when the relationships between population growth and the physical environment are considered, demography had advanced little beyond Malthusian arithmetic (Hogan 1992). Of particular importance is the need to more fully understand the nature of dynamic interactions between population—a major driving force in global environmental change—and natural resources. Changes in land use can be affected by both population growth and changes in population distribution resulting from migration flows.

Although considerable attention has been devoted to the impact of population growth on sensitive ecosystems in the developing world, the study of human impacts in the United States may be just as important because it affords particular opportunities for modeling and increased understanding. Given the projected growth rate for the country and certain regions within it, this understanding is vital in order to project environmental consequences. Between 1990 and 1995, the U.S. population grew by 1.0% per year on average, a slight increase from the average annual growth rate of 0.9% in the 1980s. However, growth patterns varied substantially by geographic region.

The U.S. population is one of the most mobile in the industrialized world; over the past three decades it has been steadily shifting from northern Frostbelt states to southern and western Sunbelt states. Between 1990 and 1995, the most rapid growth has been in the West, particularly in the mountain states of Nevada, Idaho, Arizona, Colorado, Utah, and New Mexico, where average population growth rates of 2% or more are common. If sustained, albeit unlikely, such growth rates would double the populations of certain states in just 35 years—a pace faster than many developing countries. Generally perceived as an economic benefit, short- and long-term population growth can also present tremendous environmental challenges. Moreover, responding to such challenges is complicated by rapidly shifting patterns of population growth. Between 1993 and 2020, the U.S. population is projected to climb from 258 million to 326 million. The South and West regions—areas already under environmental stress—are expected to account for 82% of the growth during this period. Most of the growth will occur in eight states, so land use issues and the ability to model human impacts on the environment are likely to become critical.

Data sets on population are available for a range of spatial scales for most regions on the Earth and are particularly reliable in the developed world. Two key questions that must be addressed in this context follow.

- How should such data sets be exploited to advance understanding of the human dimensions of changing land use/land cover?

- How compatible are the data sets with the diverse biological and environmental data sets currently available, particularly in relation to spatial units of analysis?

According to the Human Dimensions Program "...a high priority in research on the human dimensions of global environmental change must be placed on conceptual and methodological issues since, without appropriate concepts and methodologies, research cannot be undertaken" (Jacobson and Price 1991). This case study shows how new concepts emerge from methodological and analytical advances that seek to integrate remote sensing, environmental, and demographic data that target changes in U.S. population distribution and growth. It describes collaborative work with the Biodiversity Research Consortium on links between a suite of measures of human activity obtained from the 1980 and 1990 U.S. Censuses and landscape metrics that quantify spatial patterning of landscapes in ecologically relevant ways that can be tracked using remote sensing.

Methods

The primary unit of analysis used was the 640km^2 hexagon derived from the U.S. Environmental Protection Agency's Environmental Monitoring and Assessment Program (EMAP; see Kiester et al. 1993). A digital grid of 12,600 hexagons was overlaid onto the conterminous United States. Hexagons were chosen as the basic sampling unit because the distance between the centroids of any two adjacent hexagons is a constant 27km. Mapped data of landscape and habitat types were available from an analysis of Advanced Very High Resolution Radiometry (AVHRR) meteorological satellite images (NOAA) (Loveland et al. 1991). The satellite's sensor resolution is 1.1km^2.

The Loveland et al. land cover classification, derived from the AVHRR data, formed the basis for the study's landscape metrics. One hundred fifty-nine land cover classes of the scheme were aggregated into 13 coarser classes of an Anderson Level II scheme, and a final urban class was added from the Digital Chart of the World (Danko 1992). The 14 land cover types were cropland/pasture, grassland/cropland, woodland/cropland, grass-dominated, shrub-dominated rangeland, mixed grass/shrub rangeland, deciduous forest, coniferous forest, mixed deciduous/coniferous forest, water bodies, coastal wetlands, barren or sparsely vegetated land, alpine tundra, and urban areas.

Land cover characteristics were calculated for each hexagon at Anderson Level II (14 land cover classes) by summarizing the distribution of each land cover class across all 1.1km^2 AVHRR-derived pixels in each hexagon. In addition, landscape pattern metrics were computed for variables such as patch size distributions for various cover classes, shape complexity and fractal dimension, types and frequency of edges between habitat types, and measures of road abundance and total length of all major riparian systems present per hexagon (O'Connor et al. 1996, Hunsaker et al. 1994). Long-term weather data for average annual precipitation, mean January and July temperatures and annual temperature variation for the same period were derived from the Historical Climate Network Database.

One goal of the case study was to determine the extent to which population density or its near-equivalents either captures most anthropogenic interaction with the environment or is merely one facet thereof. Consequently, nine variables from the 1990 county-level census data file (U.S. Bureau of the Census, 1990) deemed most likely to capture key demographic facets and be positively linked to variation in land use/land cover patterns across the conterminous U.S. were collected and examined using principal components analysis (PCA). The list of variables follows.

- Change in population (1980-1990)

- Mean age of structure (1990)

- Metropolitan or nonmetropolitan status (1990)

- Total number of farms (1987)

- 1980 and 1990 population

- Total acreage in farms (1987)

- Total number of housing units (1990)

- Per capita income (1989)

The variables provided measures of population density and growth, surrogate measures of date and intensity of settlement, and measures of the urban or rural nature of the area. Quantified in terms of per capita income, wealth (a measure of affluence) was included because of the potential relationship to consumption patterns.

Because landscape pattern metrics were calculated for each hexagon and census variables were at the county level, the digital county level boundary file was overlaid onto the digital EMAP hexagon grid in ARC/INFO. Weighted values for each census variable per hexagon were calculated from the intersected coverages. Area weighting was used for density measures and population density weighting was used for per capita income. All census variables were appropriately normalized prior to PCA.

Results and Discussion

PCA was used to create a set of composite indices of human effects. The analysis generated variables that minimized the total residual sum of squares after fitting linear functions in all census variables across all hexagons. This process yielded two axes of interest, which were then interpreted as an index of *human settlement* in the case of PC1 and *density independent growth and settlement* in the case of PC2 (Mageean and Bartlett 1996). The first principal component accounted for 54% of the total variance and had positive major loadings on four variables—1980 population, 1990 population, wealth index, and housing density. When mapped across the 48 conterminous states (see the next figure), the PC1 scores broadly parallel the pattern of what most demographers would call population density, but the PC axis has the virtue of using information from multiple census variables and is better described as an index of human settlement. Note that darker tones in the image identify more densely populated regions.

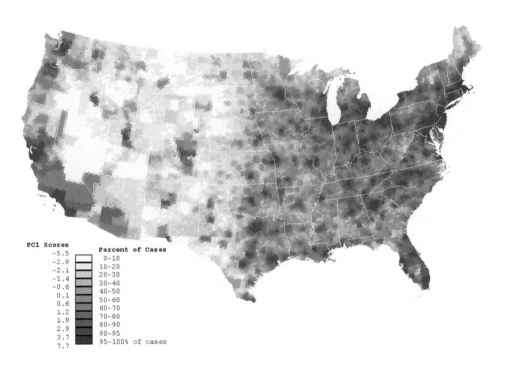

PC1 Scores
-5.5
-2.9
-2.1
-1.4
-0.6
0.1
0.6
1.2
1.9
2.9
3.7
7.7

Percent of Cases
0-10
10-20
20-30
30-40
40-50
50-60
60-70
70-80
80-90
90-95
95-100% of cases

Distribution of PCA scores for human settlement index based on nine variables extracted from the 1990 U.S. Census. Black areas denote highest values (top 5%) for the index.

From the human dimensions of the environment perspective, exploring the relationship of the index to environmental factors is worthwhile. For example, climate and topography may constrain settlement, or settlement may determine subsequent land use. Because correlation between specific landscape metrics and the study's demographic indices may be modified by other cultural, political, and/or socioeconomic variables, and because specific landscape variables may have distinct effects in different parts of the country, correlation analysis and traditional linear regression modeling are inappropriate for this analysis. Consequently, an adaptive statistical technique was used to identify significant, nonlinear, regionalized relationships among land use and climate covariates. Called classification and regression tree (CART) analysis (Breiman et al. 1984), the technique recursively partitions a focal variable (e.g., human settlement index) with respect to a set of independent variables. For each independent variable, a splitting threshold is chosen to maximize differences in the response variable (maximum between-group diversity), and the data set is split into two subsets. The independent variable that best splits the response variable explains the most variation in the data; that variable is used in the tree as a splitting variable. The process is then repeated independently and recursively on each increasingly homogenous subgroup until a stopping criterion is satisfied.

CART analysis was employed to partition the variance in the human settlement index among the environmental variables considered, and recursive partitioning yielded the hierarchical model depicted in the next image. Ovals denote split points (splitting variables are listed), while rectangles denote end points. Both ovals and rectangles contain within-group mean values for the human settlement index. The human settlement index was first split on the basis of annual precipitation with a threshold of 709mm. Drier areas followed the left-hand branch, while wetter areas followed the right-hand branch. Drier areas were subsequently segregated into nodes A and B based on seasonal differences. Wetter areas were partitioned into urban areas with greater than 2.3% of the land area classified as urban (node E) and nonurban areas. Nonurban areas were further segregated into nodes C and D based on an average July temperature threshold of 19°C.

Regression tree model rules relating an index of human settlement to environmental and remotely sensed land use variables, resulting in five end nodes or sets of hexagons with shared environmental conditions of relevance to settlement index.

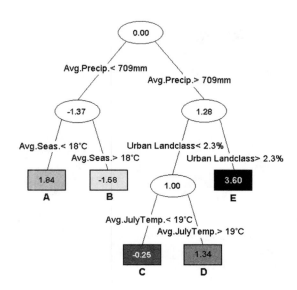

Regression tree model rules

Node	Environmental condition
Node A	Hexagons with less than 709mm annual precipitation and minimal difference between January and June temperatures (low seasonality).
Node B	Drier areas with high seasonality.
Node C	Wetter (i.e., with precipitation greater than 709mm), nonurban hexagons (less than 2.3% urban representation) with cooler summer temperatures (average July temperatures less than 19°C).
Node D	Wetter, nonurban hexagons with warmer July temperatures.
Node E	Wetter, urban hexagons with greater than 2.3% urban representation.

The next image shows the location of hexagons in each end node summarized in the table.

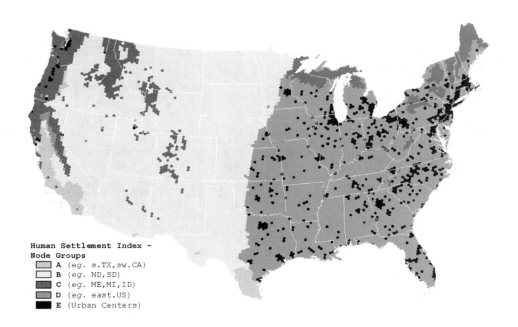

Human Settlement Index -
Node Groups
A (*eg.* s.TX,sw.CA)
B (*eg.* ND,SD)
C (*eg.* ME,MI,ID)
D (*eg.* east.US)
E (Urban Centers)

Regionalization of regression tree generated determinants for the human settlement index.

The results suggest two general conclusions. First, urban centers are primarily in the wetter East (node E, $\bar{x} = 3.60$) and are otherwise driven by geographical factors such as proximity to rivers and the Atlantic Ocean, as well as historical factors such as area of initial settlement and penetration of the country. The latter phenomenon may have been limited by aridity in the West, an idea supported by the concentration of node E sites along the Pacific Coast (see the previous image). Second, strong interactions among climatic variables appear to influence settlement patterns; in wetter, nonurban areas summer temperatures were critical (nodes C and D), with warm summers favored (node D, $\bar{x} = 1.34$) but in arid areas annual temperature variation was critical, with seasonably equable areas favored for settlement (node A, $\bar{x} = 1.84$). Note that a series of sensitivity analyses revealed no collinearity among climate variables.

While in a sense these findings are well known to geographers, this analysis allows quantification of the settlement pattern's dependence on, or independence from, environmental factors. In effect, it identifes *interactions* between settlement and environment rather than *correlations*.

Of particular interest to those concerned with the environmental impact of population is the study's second principal component. This was a multivariate structure contrasting areas of high population growth accompanied by new building with areas of farming and established settlement patterns, in essence measuring the effect of population growth and redistribution. Because this differential growth is orthogonal to the first, such differential growth is independent of the general pattern of settlement. Consequently, the index is described as density independent growth and settlement (DIGS), and it measures growth over the 1980-1990 period involving new development away from land allocated to agriculture. When analyzed in a CART tree, a node that segregated the locations with the highest values of this index was obtained. The locations of this growth were selectively concentrated on coastal barrier islands and dunes, and along the edges of desert areas, all locations of scarce fragile ecosystems. This national pattern of impact, which does not appear to have been previously documented, has major conservation implications.

While the scientific understanding of environmental and demographic change is dramatically increasing when studied separately, an ability to link the two in a synthetic and holistic way has proven elusive. Furthermore, while much of the attention surrounding the population-environment issue has been directed toward population growth, there is a need to examine the influence of other demographic processes such as migration and urbanization. This project uses GIS technology to explore new methods for evaluating the spatial relationships between population and land use. The combination of digital data, methods, and conceptual analysis incorporating remotely sensed data presented here appears to offer a powerful way of detecting human impact on the environment.

Acknowledgments

We wish to thank Biodiversity Research Consortium collaborators Carolyn Hunsaker, B. Jackson, and S. Timmons (Oak Ridge National Laboratory) for provision of landscape metrics; R. Neilsen, D. Marks, J. Chaney, C. Daly, and G. Koerper (U.S. Environmental Protection Agency Environmental Research Laboratory, Corvallis, Oregon) for assistance in computing climate data; Tom Loveland (EROS Data Center, U.S. Geological Survey) for land cover class data; and Raymond J. O'Connor (Wildlife Ecology, University of Maine) and Denis White (Geosciences, Oregon State University) for assistance with spatial analysis. John Bartlett acknowledges financial support for this work from Interagency agreements DW12935631 between USEPA and USDA Forest Service, USEPA Cooperative Agreements CR818843-01-0 and CR823806-01-0, and USDA Forest Service Cooperative Agreement PNW93-0462 with the University of Maine (Raymond J. O'Connor, principal investigator).

References

Bayfield, N.G. "Recovery of four heath communities on Cairngorm, Scotland, from disturbance by trampling." *Biological Conservation* 15 (1979), 165-79.

Breiman, L., J.H. Friedman, R.A. Olshen, and C.J. Stone. *Classification and Regression Trees.* Monterey, California: Wadsworth & Brooks, Inc., 1984.

Danko, D.M. "The digital chart of the world." *GeoInfo Systems* 2 (1992), 29-36.

Hogan, D.J. "The impact of population growth on the physical environment." *European Journal of Population* 8 (1992), 109-23.

Hunsaker, C.T., R.V. O'Neill, S.P. Timmins, B.L. Jackson, D.A. Levine, and D.J. Norton. "Sampling to Characterize Landscape Pattern." *Landscape Ecology* 9 (1994), 207-26.

Jacobson, H.K. and M.F. Price. *Human Dimensions Program: A Framework for Research on the Human Dimensions of Global Environmental Change.* Report No. 1. International Social Science Council, 1991.

Kiester, A.R., D. White, E.M. Preston, L.M. Master, T.R. Loveland, D.F. Bradford, B.A. Csuti, R.J. O'Connor, F.W. Davis, and D.M. Stoms. *Research Plan for Pilot Studies of the Biodiversity Research Consortium.* Washington, D.C.: U.S. Environmental Protection Agency, 1993.

Loveland, T.R., J.W. Merchant, D.O. Ohlen, and J.F. Brown. "Development of a land-cover characteristics database for the conterminous U.S." *Photogrammetric Engineering and Remote Sensing* 57 (1991), 1453-63.

Mageean, D.M. and J.G. Bartlett. "Putting people on the map: integrating social science data with environmental data." Proceedings of the PECORA Thirteen Conference, Sioux Falls, South Dakota, August 1996. Bethesda, Maryland: American Society of Photogrammetry and Remote Sensing, 1996.

O'Connor, R.J., M.T. Jones, D. White, C. Hunsaker, T. Loveland, B. Jones, and E. Preston. "Spatial partitioning of environmental correlates of avian biodiversity in the conterminous United States." *Biodiversity Letters* 3 (1996), 97-110.

Stern, P.C. "A Second Environmental Science: Human-Environment Interactions." *Science* 260 (1993), 1897-99.

Terborgh, J. *Where have all the birds gone?* Princeton: Princeton University Press, 1989.

U.S. Bureau of the Census. Census of Population and Housing. Summary File-3C. U.S. Department of Commerce, 1990.

Section III

Translating Applications to Social Significance

The aim of this section is to begin focusing on higher orders of resource classification and management. The 1976 Anderson classification system has long been a standard for using aerial and satellite data in management schemes, but in the absence of supporting technologies and tools, such images have been limited mostly to the lower levels of abstraction (i.e., Levels 1 and 2). By including GIS and GPS technologies, and finer ground resolution data from the next generation of satellite sensors, it is possible to get closer to the importance of economic and demographic data for resource management. As the next chapter illustrates, linking people with pixels offers the opportunity for using Anderson Levels 3 and 4 in future management scenarios.

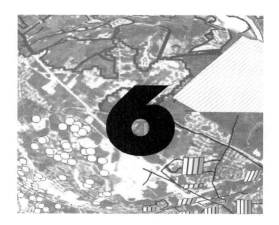

Multi-scale Economic and Demographic Data

D.J. Cowen and J.R. Jensen, University of South Carolina-Columbia

This chapter explores the transition between natural and cultural resources and the uses of remote sensing and GIS in linking people with pixels. Its aim is to demonstrate how the spatial and spectral characteristics of various sensing systems can be used to extract diverse features in a cultural setting. Data sources with spatial resolutions ranging from 20m x 20m to 2.5m x 2.5m are examined. Most analysis is accomplished within a GIS computing environment rather than traditional image processing, an approach intended to demonstrate that GIS and image processing software have evolved to a stage where boundaries between the two tools are often hard to define (Cowen 1997).

Remote Sensing of Urban/ Suburban Attributes

Landscapes created by human beings are complex mosaics of natural and manufactured materials (e.g., concrete, asphalt, plastic, shingles, water, grass, and shrubbery) arranged in organized ways to build transportation systems, utility lines, homes, commercial buildings, parks and other public spaces. Characteristics of many of these phenomena can be remotely sensed from suborbital aircraft or satellite, yielding both qualitative and quantitative information. To remotely sense phenomena, it is important to stipulate exactly what spatial resolution is required to adequately characterize attributes of interest. For example, accurate cadastral maps usually require a minimum mapping unit of from 0.3m to 5m (0.98' to 16.4'). Therefore, even the French SPOT panchromatic data which currently provide the best digital satellite ground resolution of 10m x 10m do not satisfy this data requirement.

In addition, information about urban and cultural attributes is best collected using very specific regions in the electromagnetic spectrum, as shown in the next table.

Data	EMS wavelength	Abbreviation
Land cover (U.S. Geological Survey Level III)*	Visible (0.4µm to 0.7µm) Near-infrared (0.7µm to 1.1µm) Middle infrared (1.5µm to 2.5µm)	V NIR MIR
Building perimeter, area, volume, and height	Black and white panchromatic imagery (0.5µm to 0.7µm)	Pan
Urban temperature	Thermal infrared (3µm to 12µm)	TIR
*Source: Anderson et al. 1976.		

Future sensors may include commercial ventures such as EOSAT/Space Imaging IKONOS (1m x 1m pan; 4 x 4 multispectral), EarthWatch Earlybird (3m x 3m pan), OrbView 3 (1m x 1m pan), the U.S./Russian SPIN-2 (2m x 2m), and the Indian IRS P5 (2.5m x 2.5m) (Montesano 1997). Level IV classes may best be monitored using high spatial resolution sensors including aerial photography (0.3m to 1m), and proposed EarthWatch Quickbird pan (0.8m x 0.8m) and IKONOS (1m x 1m) data. The deployment of these new satellites suggests that remotely sensed data will be available at spatial resolutions suitable for detecting features significantly smaller than those discernible in images derived from current commercial sensors. As a result, a dramatic increase in the use of remote sensing for urban applications is likely.

Computing Environments for Handling Remotely Sensed Data

Remote sensing constitutes a valuable method for collecting and transforming images of the Earth's surface. The objective of digital image processing was described by Kao in 1963 as follows: "Before the computer can process and analyze them, the images must somehow be translated into [an] analytic form acceptable to it. This is a rather difficult problem without satisfactory solution as yet and in computer programming comes under such various terms as 'artificial intelligence' and 'pattern recognition.'"

In functional terms, nothing has changed in the past 30 years. In contrast, radical change occurring in the computing environment has made handling remotely sensed data possible, and determines how data are integrated within the larger field of GIS. The GIS community tends to view remote sensing as a data input technology, while remote sensing specialists perceive GIS data as useful ancillary information to improve classification accuracies (Cowen 1988). Regardless of the point of view, the first step in any form of automated geospatial processing consists of transforming analog models of objects on the Earth's surface into machine-readable formats. Remote sensing systems use scanners to collect analog or digital data that are typically recorded in 2^6 to 2^8 digital values. The remote sensing analyst then uses a set of software tools to preprocess images in terms of radiometric and geometric corrections. During this step, brightness values are converted into geocoded images that are rectified to features on the Earth's surface and aligned to a standard map coordinate system (Jensen 1996). Next, image processing tools are used to enhance the data and extract meaningful information. Data enhancement and information extraction involve a variety of visual and statistical procedures that classify the original images into geographical features such as land cover categories.

The relationship between raster image formats and raster based GIS (i.e., gridded) data structures represents the interface between the fields of image processing and vector based GIS, and it blurs the boundary between the two approaches to managing natural and cultural resources. Without diminishing the complex issues related to processing and classifying imagery, from a GIS perspective no difference exists between a matrix of numbers generated by a scanner and other derivations of a gridded GIS layer. Represented as grid layers, bands of multispectral data can be manipulated and displayed in a variety of ways. In addition to the typical options for combining map layers via algebraic operations and enhancing features with spatial filters, it is also possible to treat brightness values as a continuous surface (Tomlin 1990, Berry 1987). In response to user demand, software vendors in the GIS market have incorporated many functions considered to be core elements in remote sensing software, and vice versa.

The most obvious area for this blending has been in raster based GIS. Distinctions between the software suites have traditionally involved integrating vector or polygon data structures. An article by Ehlers et al. (1989) offers a useful starting place to evaluate the integration of remote sensing and vector GIS technology. In brief, the authors characterized the state of the art in 1989 as one of "separate but equal" systems between which data formats could be exchanged. Toward the end of the 1980s, research was just beginning to address the conversion of image data into polygons for direct analysis with other GIS layers (Cowen et al. 1988). The rather cumbersome vectorization procedure required data conversion in a remote sensing system and then subsequent file transfer to a polygon based GIS. In other words, at the end of the last decade sophisticated users could incorporate remotely sensed data into GIS applications regardless of the data structure. However, it was not common practice. Since that time, remote sensing systems have expanded their ability to display and convert vector GIS layers.

In 1998 a "seamless integration" of the two technologies still does not exist. Vector based GIS systems do not have procedures for unsupervised classification of remotely sensed data, and remote sensing systems do not support full polygon processing. Wilkinson (1996) suggested that classification and spatial generalization remain the major issues because the process is still dependent on parametric classifiers to help model spectral feature space and develop reasonable data classifications. These routines are based on per pixel statistical pattern recognition procedures that yield a great deal of noise. In order to be useful in a GIS environment, these pixels should be spatially generalized or segmented into homogeneous polygons. He also argued that the GIS user still demands polygon based data layers that are more efficient to store than their raster counterparts. According to Wilkinson, the generalization process is critical to GIS database development, "that is, to simplify the spatial structure based on a guiding principle that the thematic map should be made as visually simple as possible but on the basis that only information from the image domain enters the process. The generalization of the resulting pixel based thematic map is a difficult problem" (Wilkinson 1996:88).

The remainder of this chapter focuses on the questions raised by Wilkinson. Using a combination of image processing and GIS tools, several forms of remotely sensed data are reviewed in terms of their utility for conducting spatial analysis and extraction of features in complex environments where landscape elements with human dimensions predominate. Only when these finer resolution elements are integrated into analysis strategies will it be possible for resource scientists to fold social and economic trends into management schemes.

Experiments with Spatial Resolution

Urban Encroachment

The first example demonstrates how 20m x 20m multispectral data by SPOT Image Inc. can be used to measure urban expansion. Using the sensor's three bands (see SPOT 1988), a team of analysts from the South Carolina Department of Natural Resources employed an unsupervised classification approach to generate state-wide Anderson Level I (Anderson et al. 1976) land cover classification in 1990. The output of this analysis was a file of 20m grid cells registered to the Universal Transverse Mercator (UTM) coordinate system. The grid cells were subsequently converted into vector polygons, which were then compared with the 1978 U.S. Geological Survey (USGS) Land Use/Land Cover (LULC) polygons (USGS 1983) that had been manually interpreted as polygons from aerial photography. By converting the SPOT image into a polygon format, it was possible to use GIS polygon overlay procedures to analyze the changes. The polygon overlay process preserved the spatial integrity of the USGS data and provided a clear way to identify areas converted to urban land cover during the 12-year period.

Expansion of urban land
cover from 1978 to 1990
for Columbia, South Carolina,
based on two data sources:
1978 urban land cover
derived from USGS data
(A), 1990 urban land cover
derived from SPOT data (B),
and changes from 1978 to
1990 (C).

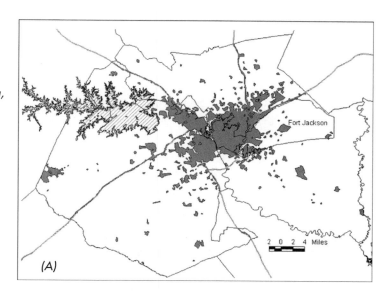

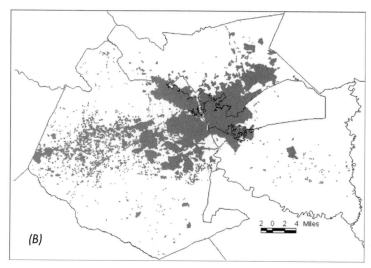

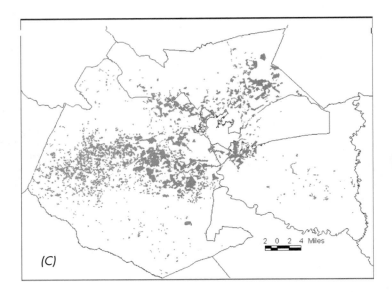

(C)

Residential Forecasting

The second set of examples analyzed the ability of SPOT 10m x 10m panchromatic data to be input to an urban forecasting model for the purpose of measuring changes in smaller sections of an urban area (Jensen et al. 1994). For a study of residential expansion, the trends in building permits in a census block group were compared to the amount of developable land. Coupled with average house size, these factors provided a method for estimating land absorption rates throughout the Columbia, South Carolina, urban area. In this analysis the SPOT panchromatic data provided a basis for determining precision limits to residential expansion. This level of detail was previously unachievable using traditional urban data sources such as housing figures from the 1990 Census.

Urban residential forecasting model based on building permits and SPOT data. SPOT 10m data with building permits (A), developable land (B), predicted trend in new homes from 1990 to 2005 (C), and predicted trend in new homes by census block group (D).

(C)

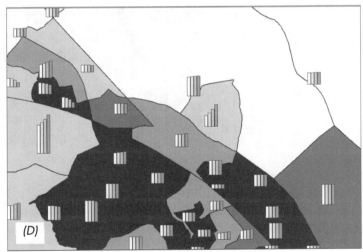

(D)

Neighborhood Change Detection

SPOT panchromatic data were also analyzed for the period spanning 1990 through 1996 to evaluate neighborhood change. In this case, county tax maps and assessor records were used as the reference in a rapidly growing part of the Columbia metropolitan area. Because the SPOT panchromatic band senses brightness values in the visible spectrum, it effectively distinguishes cleared or developed areas (increased brightness) from areas dominated by natural or cultural vegetation (decreased brightness). A comparison of the 1990 and 1996 images clearly shows that a new subdivision was constructed during the period (A and B in the next set of images).

Even more information was obtained by converting the images into grids and examining brightness differences. From a spatial analysis perspective, it is important to be able to directly convert a SPOT—or similar—image into a grid, because at that point it can be treated as a matrix of numbers. In this case the numbers corresponded to the 256 brightness values of the SPOT panchromatic band (ranging from green to red), which could be classed and symbolized using a variety of methods (e.g., equal interval, equal area, natural breaks, standard deviations, and so forth). Once the data were converted to grid values, the layers were subtracted using map algebra. Highlighting areas of substantial change was made possible by focusing only on values that were more than one standard deviation above or below the mean. The resultant "change polygons" were displayed with the 1990 SPOT data to pinpoint areas of greatest change.

Neighborhood change detection using SPOT data. 1990 SPOT image of study area (A), 1996 SPOT image (B), green areas (C), cleared areas (D), 1990 to 1996 change (cleared areas appear whiter, green areas darker) (E), and overall change (F).

Real estate professionals and local planners view this type of analysis as an excellent way to monitor change in a rapidly developing area. It enables them to detect changes that cannot be identified using customary windshield surveys. One interesting aspect of the analysis was the conversion of the SPOT images into grid cell data and then polygons. The conversion was performed using simple menu options on a desktop GIS system (ArcView spatial analyst extension). This process was considered to be quite sophisticated as recently as 1994, but is commonplace today.

Another useful form of sensor data is high resolution multispectral imagery obtained from aircraft instead of spacecraft. In a GIS processing environment, such data can accomplish the type of cartographic specification that Wilkinson (1996) seeks in GIS analysis. In order to simulate the type of imagery that can be expected from the next generation of remote sensing satellites, NASA's Stennis Space Center developed the Calibrated Airborne Multispectral Scanner (CAMS) (Jensen et al. 1994). The data have an effective spatial resolution of 2.5m or 5.0m and cover nine bands ranging from blue to thermal reflectance values (0.42 to 12.5 μm). Data from this sensor were acquired for a study area in northwestern Columbia, South Carolina. An initial graphic comparison of the spatial resolution of the 5m x 5m and 2.5m x 2.5m CAMS data with the SPOT data is provided in the next three images (Cowen et al. 1991). In this example, the Census Bureau's TIGER (Topologically Integrated Geographic Encoding and Referencing)/Line road file for the area is overlaid on the images. The SPOT data fail to provide an adequate backdrop for editing road positions, but the 2.5m resolution CAMS data adequately serve the purpose.

Comparing a TIGER road
file with remotely sensed
data collected at different
spatial resolutions. SPOT 10m
data (A), CAMS 5m data (B),
and CAMS 2.5m data (C).

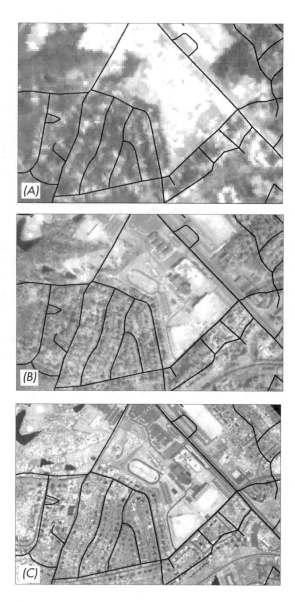

(A)

(B)

(C)

Housing Density

In order to determine the value of CAMS data at 5m resolution, an experiment was undertaken to identify specific

houses in a residential subdivision (Cowen et al. 1993). The experiment counted houses in 32 census blocks. The number of houses was determined using photo identification with a 1:4,800 orthophoto and field verification. The CAMS data were classified via an unsupervised procedure using the near IR, red, and thermal infrared bands, and houses that appeared on the image were converted into polygons. Eliminating polygons smaller than 25 mi^2 strongly correlated (.94) with the actual number of houses in the study area. The results suggest that multispectral data, especially with a thermal band, provide a useful estimate of home counts for a given residential area.

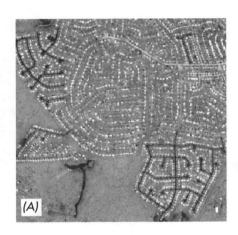

Housing counts based on CAMS 5m x 5m data. Raw 5m data (A), houses classified from the CAMS thermal band (B), and houses classified from the resulting orthophoto (C).

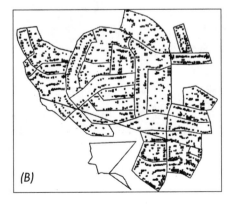

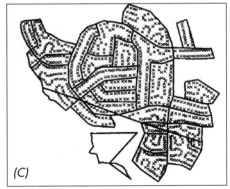

Extracting Houses and Roads

Visual inspection of the previous CAMS image suggests that 5m CAMS data might be particularly useful where little tree canopy interferes. The next series of images represents loadings on the first principal component formed from the nine original CAMS bands. In other words, it represents a statistical summary of several intercorrelated bands. Nevertheless, within a gridded GIS environment these values are simply another layer or matrix of numbers. Grid values can be easily manipulated with a legend editor to reveal basic house and road themes. Because the numbers represent continuous values, they can be displayed as contours of equal value ranges. This representation clearly outlines house and road features and helps determine the proper range of values to use in order to classify the grid into different themes.

Once ranges are delimited, the grid can be reclassified into binary maps of houses and roads and then converted into polygon themes. The entire process—converting the image to a grid, manipulating the grid, and reclassifying grid values into polygons—is handled with spatial analysis tools in a GIS environment. In fact, the road layer was further processed with a neighborhood majority filter that eliminated small clusters of cells not contiguous with cells representing roads. This procedure indicates that in areas with little interference from overhanging trees, 5m spatial resolution data can be used to detect specific urban features such as houses and roads. Recreation facilities in forests, national parks, national monuments, and state managed facilities might similarly be assessed and inventoried.

Extracting house and road layers from CAMS 5m x 5m data. Raw 5m data (A), classified grid (B), value contours (C), road layer (D), house layer (E), and both layers overlaid (F).

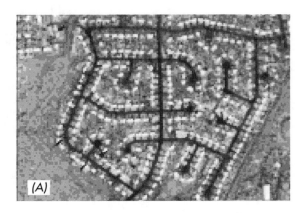

(A)

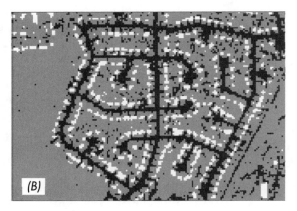

(B)

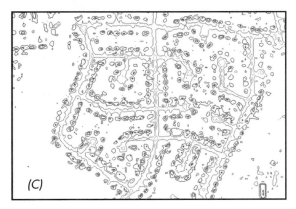

(C)

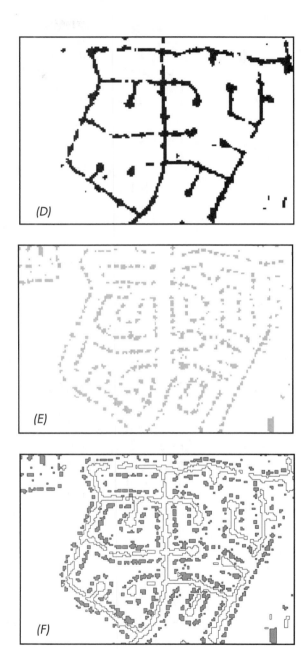

(D)

(E)

(F)

Urban Recreational Features

The final experiment involved identifying urban features from the CAMS data at a 2.5m spatial resolution. In this case different bands were used to extract a baseball diamond, tennis courts, and a general category of cultural features that includes houses, roads, and buildings. The analysis involved selection of the best band for each feature. (See the final series of images.) The green band (0.52 to 0.60 μm) effectively separates dirt and grass on the baseball diamond. The 90' base paths effectively illustrate the spatial characteristics of the 2.5m data.

By focusing on differences between the grass and infield soil, it is also possible to separate grass from trees. By contrast, the green band does not provide sufficient information for identifying the track or other cultural features. The color differences of the tennis court surface are clearly evident on the CAMS near infrared band (0.76 to 0.90 μm). In fact, this combination of spatial and spectral characteristics provides evidence that identifying features as small as a tennis court (36' x 78') is possible with 2.5m data. The general classes of cultural features (e.g., roads, houses, track, and so forth) are readily apparent on the thermal infrared band. Conversely, this band (which measures heat emanating from the surface) is not useful for separating grass from trees or colors on the tennis court. The results of this procedure show the value of multispectral data and indicate that remote sensing at spatial resolutions of less than 5m provides valuable information for linking coarse-grained natural features with interspersed, finer grained human elements.

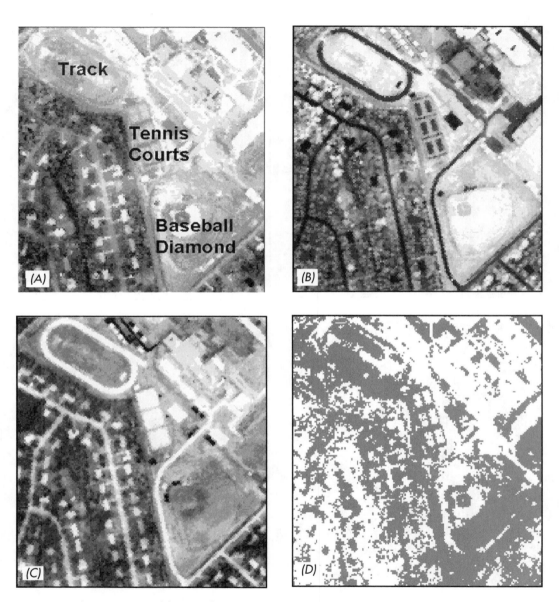

Extracting urban recreational features from CAMS 2.5m data. The green band (A), near IR band (B), thermal band (C), grid derived from the green band (D), grid derived from the near IR band (E), grid derived from the thermal band (F), the baseball diamond polygon (G), tennis court polygon (H), and cultural feature polygon (I).

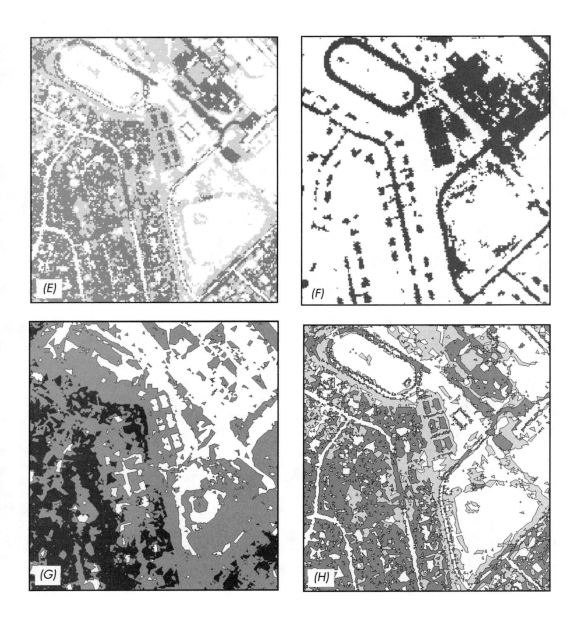

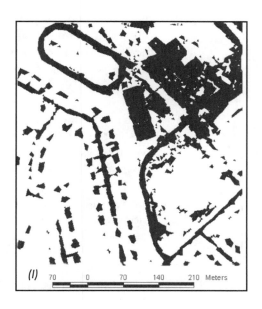

(I) 70   0   70   140   210 Meters

Conclusions

An important objective of this chapter was to review the status of integrating GIS and remote sensing technologies for assessing the interface between natural and cultural landscapes. Linking natural resource management with the human dimension of cultural resources generally requires finer resolution data. To make the evaluation, the capabilities of one type of desktop GIS software to display and manipulate remotely sensed data were examined. The results indicate that many functions necessary for handling digital imagery now exist in user-friendly systems that run on Intel based computers under standard operating systems. Imagery can be directly displayed and easily converted into both grid and polygon data structures. These capabilities suggest that from a data structure point of view there is a close coupling of GIS and remote sensing tools.

It is also clear that when image data are converted into a grid data structure, a wide range of available tools can help display, segment, separate, and generalize geographic features. Furthermore, as a result of stiff competition in both the PC based hardware and software markets, it is clear that GIS tools capable of incorporating remotely sensed data are now accessible to even small firms, public agencies, and schools. In a real sense, major technological and financial constraints that previously limited use of related technologies in the 1980s have disappeared. The next phase of this evolution is likely to be the widespread conversion of the most sophisticated GIS and remote sensing systems from the UNIX operating system to Windows NT. This implies that by the end of the decade a new breed of inexpensive GIS and remote sensing systems capable of running applications dynamically linked throughout the desktop is likely. These spatial data handling operations are also likely to be dynamically linked to applications that retrieve data and perform functions via the Internet.

In the future the GIS community will increasingly rely on remotely sensed data for timely, accurate management updates. Conversely, the remote sensing community will recognize that GIS represents a logical market for its products. In effect, this indicates that even small organizations with limited resources can perform rudimentary remote sensing tasks to supplement traditional windshield surveys. Several remote sensing organizations are currently providing data preprocessed into an image format directly compatible with GIS. All image rectification and registration procedures are essentially being handled by service or data providers, thereby eliminating much of the need for certain image processing functions. As the decade progresses, these two technologies will likely fuse into a more general category of spatial data handling systems that feed models and spatial decision support systems. Therefore, it is important that researchers in each field combine talents and concentrate on the development of efficient procedures to "classify," store, and analyze the tremendous volumes of remotely sensed data that will soon be transmitted by the next generation of satellites.

Another objective of this chapter was to offer a preview of the type of high resolution multispectral, satellite based imagery that can be expected in the near future. Examining several applications illustrated that remotely sensed data with a spatial resolution of 20m or greater is a worthwhile source of natural and cultural land covers. SPOT 10m data provide a useful source of information for monitoring relatively small changes at the urban fringe. With spatial resolutions smaller than 10m, it is possible to identify specific cultural features. CAMS 2.5m aircraft data provide useful insights regarding applications of the next generation of commercial remote sensing platforms linking human and natural elements in a variety of social and economic landscapes. The results of each experiment are extremely promising. Developers and planners should be able to use the data to periodically monitor change on an annual rather than decennial basis.

References

Anderson, J.R., E.E. Hardy, J.T. Roach, and R.E. Witmer R.E. *A Land Use and Land Cover Classification System for Use with Remote Sensor Data*. U.S. Geological Survey Professional Paper 964. Washington, D.C.: USGS, 1976.

Berry, J K. "Fundamental Operations in Computer Assisted Map Analysis." *International Journal of Geographical Information Systems* 1:2 (1987), 119-36.

Cowen, D. J., J.R. Jensen, and W. Smith. "Integration of Thematic Mapper and DLG Data for Timber Stand Assessment." Proceedings, Third International Symposium on Spatial Data Handling, IGU Commission on GIS, pp. 39-55. Sydney, Australia, 1988.

Cowen, D.J. "GIS vs CAD vs DBMS." *Photogrammetric Engineering & Remote Sensing* 54:11 (1988), 1-55.

Cowen D.J., J. Jensen, and J. Halls. "Maintenance of TIGER Files Using Remotely Sensed Data." *Proceedings, ASPRS* 4 (1991), 31-40.

Cowen, D.J., J.R. Jensen, M. King, and J. Halls. "Estimating Housing Density with CAMS Remotely Sensed Data." *Proceedings ASPRS* 2 (1993), 35-43.

Cowen, D.J. "The Practical Integration of Remote Sensing and GIS." Proceedings of an International Workshop on New Developments in GIS, pp. 1-16. International Society for Photogrammetry and Remote Sensing, ERIM, 1997.

Ehlers, M., G. Edwards and Y. Bedar. "Integration of Remote Sensing and Geographic Information Systems: A Necessary Evolution," *Photogrammetric Engineering & Remote Sensing* 55:11 (1989), 1619-27.

Jensen, J.R., D. Cowen, D.J. Halls, J. Narumalani, S. Schmidt, N. Davis, and B. Burgess. "Improved Urban Infrastructure Mapping and Forecasting for BellSouth Using Remote Sensing and GIS Technology." *Photogrammetric Engineering and Remote Sensing* 60:3 (1994), 339-46.

Jensen, J.R. *Introductory Digital Image Processing: A Remote Sensing Perspective,* 2nd ed. Upper Saddle River, New Jersey: Prentice Hall, 1996.

Kao, R.C. "The Use of Computers in the Processing and Analysis of Geographic Information." *Geographical Review* 53 (1963), 530-47.

Montesano, A.P. "Roadmap to the Future." *Earth Observation* 6:2 (1997), 16-19.

D. Tomlin. *Geographic Information Systems and Cartographic Modeling.* Englewood Cliffs: Prentice Hall, 1990.

SPOT Image Inc. *SPOT User's Handbook,* 2 vols. Reston, Virginia: SPOT Image Co., 1988.

U.S. Geological Survey. "Land Use and Land Cover Digital Data." USGS Circular 895-E. Washington, D.C.: U.S. Government Printing Office, 1983.

Wilkinson, G.G. "A Review of Current Issues in the Integration of GIS and Remote Sensing Data." *International Journal of Geographical Information Systems* 10:1 (1996), 85-101.

Coping Strategies in the Sahel and Horn of Africa: A Conceptual Model Based on Cultural Behavior and Satellite Sensor Data

S.D. Prince, M.E. Geores, and J. Boberg, University of Maryland

Resource Management Requirement

Socioeconomic data on human activity are most often collected on either the national or household scale, which are too coarse or fine, respectively, to readily interface with biophysical data available from satellites. National scale data on population, macroeconomic indicators, and agricultural production are useful for establishing a context for evaluating the coping strategy decisions made by communities, but do not explain them. In developing countries in particular, little research and data collection have been focused on scales in between.

Most recent work on the lifestyles of human populations in the Sahel and Horn of Africa has been stimulated by famines and attempts to establish famine early warning systems. Integrating data from physical, biological, and social systems is critical in order to create effective warning systems. Biophysical data from satellite measurements on precipitation and vegetation are important components of existing warning systems in the region. These data are georeferenced and available in various GIS formats. While such data are necessary for an effective warning system, they are insufficient because they do not take into consideration the availability of food for consumers affected by food supply crises.

Social Science Approaches

Social scientists have approached famine from the perspective of human behavior, particularly that of vulnerable groups. Sen's "entitlement theory" of food crises (Sen 1990) considers the uneven entitlement (or access) to food that leads to starvation in the face of potentially adequate food supplies. Sen outlines four types of entitlement: trade based, production based, own labor based, and inheritance (transfer based) access. The physical and biological triggers of famine are "state variables," or unvarying conditions under which the phenomenon of interest is played out. While famine is the phenomenon being studied, decisions are made and carried out at the household level. In addition to differences among entitlements, household coping strategies also differ when a food shortage exists. These strategies range from relatively innocuous responses (e.g., reducing consumption, temporary migration of young men for wage labor, and reliance on extended families) to major decisions (e.g., sale or abandonment of productive assets and permanent migration) (Watts 1983).

Earth System Science Approaches

Physical and biological aspects of Earth system science consider relationships that operate from regional to global scales. Studies at these scales have received an enormous boost from the advent of global observations from Earth-orbiting satellites. For the first time, it is possible to obtain direct, repetitive, and spatially comprehensive observations of the Earth's surface, instead of relying on only point measurements at ground stations. Satellite remote sensing has improved the spatial resolution and variety of variables that can be observed but in no case has the technology replaced ground observations. In the context of natural resource management, the most useful new measurements relate to land cover and land use, vegetation production, and energy balance modeling.

Prompted by inquiries made possible through satellite data, human dimensions of Earth system science are being approached in a new way. Nevertheless, until very recently the emphasis in social and cultural research has tended to be at local scales or has not been georeferenced at all. This state of the research is presaged by the fact that the search for greater understanding traditionally requires a reductionist approach. The explanation of a phenomenon, some say, is to be found at the next lower level in the hierarchy of complexity, while factors at the next higher level determine the state variables. Thus, there is a natural trend toward finer detail and, hence, less relevance to broader scale issues.

The same may be true regarding the biological aspects of Earth system science. Having posed problems at the scale of human experience (e.g., spatial scales ranging from several hectares to several square kilometers, and temporal scales ranging from several days to several years), ecology has moved into more and more detailed explanations in space and time. As a consequence, the dramatic blossoming of observation systems that make measurements at scales from 20m to 5km has yet to be fully exploited. The same is true of the human dimensions of global change and Earth system science; hence, the need to recognize that studies at coarser scales—from administrative districts to continents, and even the entire land surface of the Earth—are based on the same, firm philosophical foundation as studies at finer resolutions.

Both the biophysical and human sciences contain exceptions to these generalizations. Meteorology and oceanography have traditionally been more concerned with regional and global scales, because the processes under study embrace broad geographic areas. In human sciences, economics and demographic studies have typically been applied at a coarser spatial scale than sociological and cultural studies, and have appropriate measurement techniques available to them (e.g., census and macroeconomic data provided by trade statistics).

This case study explores relationships between the environment of the Sahel and Horn of Africa, and the coping strategies of human populations in the region who depend on agriculture and pastoralism. Although in its early stages, this study indicates directions and data requirements that may apply to similar studies probing relationships between natural resources and humans at regional and continental scales.

Environmental Limits of the Study Area

The Sahel is the name given to the transitional area between the southern reaches of the Sahara Desert and savanna regions to the south. It extends from the coast of the Atlantic Ocean in Senegal and Mauritania in the west, to the Red Sea coast of Sudan in the east, a distance of 5,500km. Conservatively, the Sahel covers an area of 1.5×10^6 km^2. One of the largest biomes on Earth, the northern and southern boundaries of the Sahel are among the most dynamic of any on the globe. The zone they define is easily visible from orbiting spacecraft (item A in the next illustration). The name "Sahel," derived from the Arabic word for "coastline," describes the place where vegetated and barren landscapes meet.

Mean seasonal sum of NDVI (normalized difference vegetation index) for the 1981-1989 period for Africa north of the equator, indicating mean primary production (A). Interannual standard deviation of mean growing season NDVI for the same period, showing the "between year" variability in primary production (B). Coefficient of variation (standard deviation/mean) of NDVI for the same period (C).

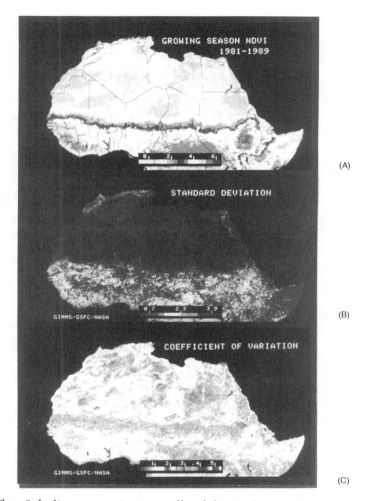

The Sahelian zone is normally delimited according to its mean annual rainfall. A typical definition of the Sahel is a bioclimatic zone of predominantly annual grasses with some shrubs and trees receiving a mean annual rainfall from 150 to 600mm. Rainfall drops off sharply in the zone's northern regions, accompanied by an increase in interannual and spatial variability of rainfall (Nicholson et al. 1998).

The desert zone in the north is mostly barren except in mountainous areas and along drainage lines, where orographic rainfall supports shrubs. Irregular flushes of vegetation associated with sporadic rainfall provide grazing for livestock and habitat for desert locusts in their recession areas. Moving southward into the Sahel, vegetation changes from scattered bushes just south of the desert margin to an increasing cover of annual grasses with a higher density of bushes and occasional trees in the center of the zone. Toward the southern margin with the Sudanian zone, perennial grasses, scattered trees, and patches of trees with bushes can be found. Most of all, the Sahel is marked by interannual variations in species composition and biomass resulting from uncertain rainfall. Because of high rainfall variability, the demarcations among the three zones are complicated by southerly components appearing in dryer, northern areas, and vice versa.

By contrast, the Sudanian zone is distinguished by its higher rainfall (exceeding 600mm); greater biomass comprised of taller, perennial grasses, shrubs, and trees; and park-like, savanna physiognomy. Along with these broad transitions are many changes in species, phenology, morphology, and physiology.

The Sahel is also a zone of cultural transition, where Islamic cultures from the north mingle with traditional cultures of the south. The agricultural systems found in the Sahel are mainly livestock herding and sedentary agriculture revolving around the availability of natural resources. Maintaining a balance between livestock and household size is an important concern, especially for pastoralists (see Thebaud 1995). Vegetation production has a strong influence on critical household decisions, such as whether to migrate on a seasonal or permanent basis, or whether to abandon livestock herding for a more sedentary, agricultural existence. These choices are examples of coping mechanisms used in times of famine. A rigidly applied subsistence taxonomy, separating social systems into agriculture, pastoralism, and foraging is no longer an adequate way to classify subsistence cultures (see Hitchcock et al. 1989). Households employ aspects of all systems in times of crisis (Graef 1997). As with vegetation, there is some stratification of social systems between the northern and southern regions, with more northerly cultures tending toward pastoralism and more southerly cultures practicing sedentary agriculture.

The southern and northern boundaries of the Sahel are defined arbitrarily. It has been argued that the northern boundary is indicated by the absence of rainfall adequate for plant life. However, if this definition is adopted, spatial and temporal variations in rainfall would cause the boundary to shift. Another definition adopts a low threshold of average annual rainfall but this also presents problems because soil moisture, not rainfall, is relevant. Using average rainfall amount is questionable when the interannual variation is strong. Therefore, we argue that the variation in vegetation production caused by varying rainfall is a more useful way to locate the northern boundary (see Tucker et al. 1991b) because this circumstance determines viable plant physiologies and human subsistence lifestyles.

The same is also true of the southern boundary. If defined by average annual rainfall, the southern boundary is located arbitrarily. Judged by species or annual production, there is no clear boundary between vegetation in the Sudanian and Sahelian zones. Hence, temporal variations in primary production may again provide a more functional definition.

In the Horn of Africa, sedentary farmers reside only in the Mogadishhu-Kismayo-Dollo area of Somalia and Jigjiga in Ogaden. The balance of the region is largely desert similar to the Sahara. Conditions in these farmed areas correspond to the Sahel in terms of production, interannual variability, and human populations.

Remotely Sensed Measurements of Primary Production

The importance of spatial and temporal intra- and interannual variation in rainfall and vegetation in the Sahel suggests that both high temporal frequency and large area coverage are needed to characterize and visualize the environment. Satellite observations are the only practical way to acquire these data in order to make measurements. They can be made at sufficiently high temporal frequencies to detect vegetation response to individual rainfall events, and they cover sufficiently large areas to capture spatial variations (Prince 1991). Satellite observations of reflected red and near infrared radiation from the land surface indicate the absorption of radiation by vegetation, and the two reflectance measurements are combined to create a normalized difference vegetation index (NDVI). Temporal integrals of these data (ΣNDVI.days) are related to primary production.

Because satellite data have been available since 1981, it is possible to calculate annual totals of ΣNDVI.days and compare them to determine interannual variability. The mean annual production (item A in the preceding figure) and its standard deviation (item B) can further be combined to show the interannual coefficient of variation, used here to illustrate the Sahelian zone and southern Horn of Africa (item C above). The areas of greatest interannual variation in vegetation in Africa are at the edges of the deserts, where significant movements of the isolines of specific vegetation indices occur from year to year (Tucker et al. 1991a).

The high coefficient of variation defines the biological character of the zone. High interannual variation (or unpredictability of environment) imposes constraints on the biota that lead to distinct ecosystem forms and functions. In more extreme environments, primary production is predictably low because the environment is predictably unsuitable for plant growth. In favorable environments, primary production is generally high and relatively invariant from year to year, although highly seasonal. In the marginal zones, where environmental suitability for growth cannot be predicted from one year to the next, plants, animals, and humans must adapt. This functional definition of semiarid lands conveys significant information about the character of these ecosystems beyond their average climate limitations (Galvin and Ellis 1996).

Human Responses to Primary Production and Interannual Variability

Given the spatial and interannual variability in primary production, a model of interaction between human and environmental factors has been developed to characterize coping strategies. North of the Sahel and in the desert regions of the Horn, low productivity leads to a low carrying capacity for animals and limited opportunities for cultivation. South of the Sahel, and to a lesser extent in southern Somalia, primary production increases and is subject to much less variability. At the steepest part of the mean annual production gradient, a zone with very high temporal and spatial variability of annual production exists, a phenomenon that decreases on either side of this point (i.e., lower and higher sides of the production gradient). These two gradients may have great significance for humans and natural biota.

One conceptual model compares annual mean productivity and its standard deviation to define four extremes: high and low production, and high and low absolute variability (refer to the following illustration). High standard deviation of production occurs in the Sahel and southern Somalia. These conditions lead to three potential human coping strategies: sedentary farming, nomadism, and transhumance (i.e., opportunistic exploitation of unpredictable resources). Sedentary farming and pastoralism are only possible above a minimum threshold of mean annual production and below a threshold of interannual variability (bottom right). At lower production, two lifestyles are possible depending on the degree of variability. At low variability (bottom left) when low production is a dependable condition, nomadic movement between small but dependable sources of resources is expected; at high variability (top left), transhumance is expected. The Fulani (or Perle) people represent an ethnic group that practices transhumance in the Sahel. The dry season is spent toward the south in grazing lands, but during the wet season the populations migrate north-ward into marginal lands where they cultivate millet and herd animals. Although the following four-cell matrix is a simple model, its utility can be tested in many regions because satellite data are acquired globally.

A conceptual model of the response of human populations in desert and semiarid savanna regions to spatial and interannual variations in primary production.

		Annual mean primary production	
		Low	High
Standard deviation of annual primary production	High	Sahel, southern Somalia Transhumance	Absent None
	Low	Desert Nomadism	Sudanian savanna Sedentary cultivation and pastoralism

It has been argued that the impact of resource unpredictability is different depending on mean production levels. At high mean production, the response to high variability could be to moderate demand to a level sufficient to sustain life and ensure survival in poor years—but not fully exploit production in unpredictably good years. However, at low mean production this option is not viable because reducing demand will not sustain the population. Mobility becomes the means to meet minimum needs. Thus, the coefficient of variation (variation in annual production normalized by the mean) may be a sensitive indicator of human coping strategies.

The proportional area under millet (item A in the next figure), a subsistence cereal having the shortest growing period of any tropical cereal, is skewed almost imperceptibly to areas of lower primary production, when compared with all crop areas (item B). Considering cash crops only (items C and D), a very strong shift of millet to areas with lower ΣNDVI.days occurs. The area of cereal cultivation, including millet, increases slightly as interannual variability of primary production increases (item E). This pattern is consistent with the fact that millet is selected for short and unpredictable growing conditions. Selection of crops for subsistence and cash is rarely determined by individuals. Presumably, crops are selected over several generations based on production performance.

An analysis of crop production (ΣNDVI.days) in the Sahel (Sebastian 1997). Proportions of land area (including noncropped) in each primary production class for millet (A) and all crops (B), and proportion of cropped area in each primary production class for millet (C) and all cash crops (D). Relationship between percentage of cropped area under millet (E) and all cereals (F) with interannual variability of production.

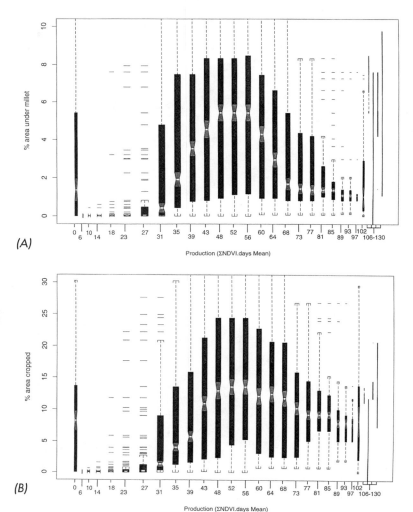

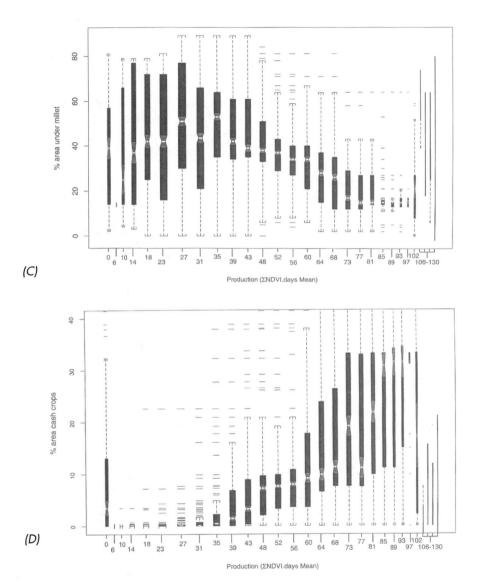

(C)

(D)

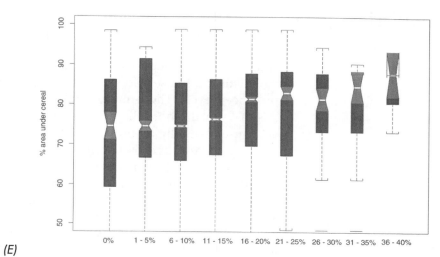

(E)

Interannual variability (coefficient of variation)

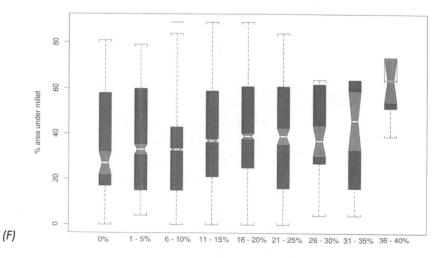

(F)

Interannual variability (coefficient of variation)

Conclusions

The experience with famines in the Sahel and Horn of Africa has demonstrated that neither human nor physical data are adequate alone to model and predict the effect of drought. Human systems need to be monitored on several scales, including the household, community, and state (district or province) levels. There are examples of ways to accomplish this task. At the household level, demographic health surveys use carefully chosen sample areas to obtain data through surveys and focus groups, and the world fertility survey uses similar techniques. Household level data should include demographic information, access to food, patterns of agriculture, ownership of animals, and other information known to play a role in coping strategies. At a community level, data can be obtained about the group's collective ability to withstand a food crisis and the social structure's organization. Ideally, baseline surveys would be undertaken in representative households and communities before a food crisis begins. Economic and social behavior also could be monitored as an early warning system to alert governments and international agencies when coping strategies start to change. At the state level, economic and social data specify the context within which households and communities are operating. These coarser scale data are closer to those used for regional biophysical monitoring. Household data can be understood as being nested within coarser scale human and biophysical data.

There are several ways to approach famine early warning with GIS. Climate and vegetation data are already georeferenced, and country—and even state or district boundaries—are readily available for use in a GIS. Gridded world population data disaggregated to the census district level are available (Tobler et al. 1995), allowing examination of the spatial distribution of population at levels smaller than political units. Country level socioeconomic data can be correlated with census data to identify where the most vulnerable populations exist using, for example, vulnerability criteria outlined by Sen. These population and socioeconomic data can be overlaid on physical and biological data to identify vulnerable regions from physical and social perspectives.

References

Galvin, K. and J. Ellis. "Climate Patterns and Human Socio-ecological Strategies in the Rangelands of Sub-Saharan Africa." In *Global Change and Subsistence Rangelands in Southern Africa: The Impacts of Climate Variability and Resource Access on Rural Livelihoods,* pp. 57-62. Workshop Report, June 10-14, 1996, Gaborone, Botswana. Working Document No. 19, International Geosphere-Biosphere Core Project, Global Change and Terrestrial Ecosystems. Canberra, Australia, 1996.

Graef, J. "Risk Minimization Activities among Vulnerable Households During the 1970s and 1980s Sahel and Horn Food Crises: Building a Coping Strategy-based Early Warning System Model." M.A. thesis. University of Maryland, 1997.

Hitchcock, R.K., J.I. Ebert, and R.G. Morgan. "Drought, Drought Relief, and Dependency Among the Basarwa of Botswana." In R. Huss-Ashmore and S.H. Katz, *African Food Systems in Crisis*. Part One: Microsystems, pp. 303-06. New York: Gordon and Breach Science Publishers, 1989.

Nicholson, S.E., C.J. Tucker, and M.B. Ba. "Desertification, Drought, and Surface Vegetation: An Example from the West African Sahel." *Bulletin of the American Meteorological Society,* 1998 (forthcoming).

Prince, S.D. "Satellite Remote Sensing of Primary Production: Comparison of Results for Sahelian Grasslands, 1981-1988." *International Journal of Remote Sensing* 12 (1991), 1301-11.

Sebastian, K.L. "The Integration of Human and Environmental Systems in the Sahel Region of West Africa." M.A. thesis. University of Maryland, 1997.

Sen, A. "Food, Economics, and Entitlements." In J. Freze and A. Sen, eds., *The Political Economy of Hunger,* Chapter 2. Oxford: Clarendon Press, 1990. Originally published in *Lloyd Bank Review* 160 (1986), 1-20.

Thebaud, B. "Human Demography and Animal Demography in Pastoral Societies of the Sahel: Towards a Better Understanding of the Pastoral Economy." In A. Mafeje and S. Radwan, eds. *Economic and Demographic Change in Africa,* pp. 36-46. Oxford: Clarendon Press, 1995.

Tobler, W., U. Deichman, J. Gottsegen, and K. Malloy. "The Global Demography Project Technical Report 1995-6." Santa Barbara, California: National Center for Geographic Information and Analysis, 1995.

Tucker, C.J., H.E. Dregne, and W.W. Newcomb. "Expansion and Contraction of the Sahara Desert from 1980 to 1990." *Science* 253 (1991a), 299-301.

Tucker, C.J., W.W. Newcomb, S.O. Los, and S.D. Prince. "Mean and Inter-annual Variation of Growing-season Normalized Difference Vegetation Index for the Sahel 1981-1989." *International Journal of Remote Sensing* 12 (1991b), 1133-35.

Watts, M. *Silent Violence: Food, Famine, and Peasantry in Northern Nigeria.* Berkeley: University of California Press, 1983.

Landscape Characterization through Remote Sensing, GIS, and Population Surveys

S.J. Walsh, B. Entwisle, and R.R. Rindfuss,
University of North Carolina-Chapel Hill

Resource Management Requirement

A major challenge facing resource managers and the broader remote sensing and GIS community is to link people to pixels at the appropriate spatial and temporal scales, so that the behavior of individuals, households, and communities can be linked to (1) changes in land use and land cover, and (2) the demographic, biophysical, and geographical processes that define the landscape in terms of composition and spatial organization. This case study describes how satellite time series data are joined with environmental gradients and longitudinal population surveys of individuals, households, and communities in northeastern Thailand. These linkages are performed within a GIS to examine behavioral, geographical, and environmental hypotheses about population-environment interactions. The study also describes efforts to explore the role of GIS in linking population, biophysical, and geographical domains, as well as approaches for characterizing the landscape through remote sensing, GIS, and population surveys.

Population and environment interactions are often perceived as unidirectional, with people forcing change on the environment. However, important environmental factors such as terrain and climate can affect human behavior, and some human induced changes in the environment (e.g., reduced soil fertility through poor management or overuse) can in turn result in feedbacks on human behavior.

Combining remote sensing, GIS, population surveys, and spatial analysis synergizes the contributions each field brings to an understanding of complex landscapes. *Remote sensing* offers an approach for characterizing the environment through measures of landscape state (e.g., plant biomass) and land use/land cover condition; *population surveys* are used to assess the behavior of people and measure their demographic and socioeconomic characteristics; and *GIS* offers a set of tools and techniques for assessing geographic location and the spatial linkages between features of similar and distinct attributes within a context extending across space and time.

The study of human behavior as it relates to environmental endowments, site suitability, and exogenous environmental effects is a relatively unexplored area within the field of resource management. The challenge is to articulate questions within an appropriate space-time framework for which disparate data sources across biophysical, social, and geographic domains exist. These data must then be organized in a compatible format for analysis involving locational and non-locational characteristics that draw upon the graphic and quantitative tools and techniques afforded through GIS and non-GIS statistical procedures.

In human terms, the hypothesis examined is that the composition and spatial organization of land use/land cover (as indicated through class and landscape metrics) are associated with biophysical, geographic, and social variables that relate, respectively, to site suitability for agriculture, village competition and accessibility to transportation and surface water, and population characteristics. Composition and pattern metrics are used to understand the form and function of landscape produced through complex population-environment interactions, which are in turn used to assess the relationships between deforestation, agricultural extensification, and population migration.

Study Area

The study area for this project is the Nang Rong district of Buriram province in northeast Thailand. The cultural environment is dominated by wet rice cultivation in the shallow depressions, with dry dipterocarp savanna forest and drought resistant crops (i.e., cassava, kenaf, jute, and sugarcane) on the uplands. The physical environment is marginal in a number of ways. Water availability and soil fertility are the critical limiting factors for agricultural production. Almost 80 percent of the total annual precipitation falls within a five-month period from May to September but the interannual amount also varies considerably. Soils are generally highly weathered and leached, are deficient in nutrients, exhibit a low cation exchange capacity, possess a low water holding capacity, and have extremely low organic matter content. The area is dominated by fine sandy loams that form an extensive area of one of the most infertile soil groups in Thailand.

Study area of Nang Rong district, Thailand.

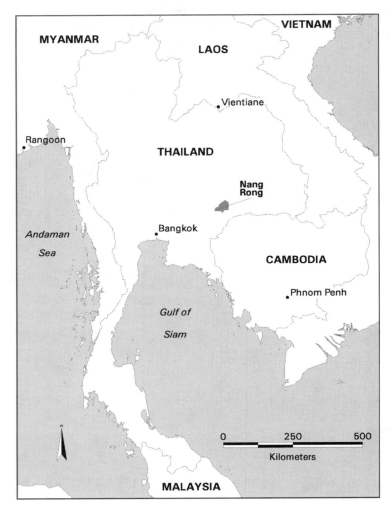

Beginning in the 1960s, many forest lands were converted to cash crop production; agricultural intensification occurred in the alluvial plains and low terraces; and agricultural extensification occurred in the uplands and on the upper middle terraces. Use of marginal lands to enhance or expand agricultural production intensified deforestation into the 1990s. Over the past 35 to 40 years, land conversion has removed nearly all natural forest vegetation and replaced it with paddy rice. On the upper and middle terraces, forest vegetation has been substantially modified by growing upland cash crops, particularly cassava.

Data and Methodology

Land Cover Changes and GIS Development

Population-environment interactions have been examined by linking population survey data collected at the household and community levels with land cover data derived from satellite spectral classifications. Additional land cover data were collected from terrain and associated landscape information digitized from base maps, and all data were integrated into a GIS. Coverages were created for villages, hydrography, topography, and the transportation network. The coverages were then converged to create views representing village territories, hydrographic and transportation accessibility, village density and competition, and topographic and resource potential.

The following sections briefly describe the five variables used in the study.

- The nature of the population surveys conducted in 1984, 1988, and 1994.

- Delineation of boundaries around nuclear villages to link data collected at discrete versus continuous surfaces.

- Derivation of topographic surfaces from generated digital elevation models (DEMs).

- Use of pattern metrics to characterize the composition and spatial organization of land use and land cover types derived from classified satellite data.

- Characterization of geographic accessibility to water and roads, as well as competition for land derived through GIS manipulations of base data.

Population Survey

Longitudinal data collection began as a community based integrated rural development project. To evaluate the success of the initial four-year project, the Institute for Population and Social Research at Mahiodol University in Bangkok collected a complete household census for 51 villages in 1984. This was followed by an enumeration of households containing a reproductive age woman in 1988. Community profiles were collected in all 51 villages in both years.

Villages, households, and individuals were surveyed again in 1994 with an expanded research agenda. In 1994, a community profile was conducted in all 310 villages in Nang Rong. The 1994 household census was conducted in the 51 villages taking part in the 1984 survey. For all households, the 1994 survey collected information regarding the whereabouts and current characteristics of household members identified in 1984. It also noted new members, visits and exchanges with former household members, annual life history data for individuals ages 18 to 35, and household characteristics such as plots of land owned and/or rented, agricultural equipment, and crop mix. In 1995, household members from 22 of the original 51 villages surveyed in 1984 were traced to selected migration destinations.

Village Boundaries

The population surveys were spatially referenced to digitized village locations and areally referenced to village territories. Village boundaries around the nuclear settlement patterns were set through alternate boundary definitions and assessed through sensitivity analyses to discern their impact on population-environment studies within the district. Ongoing spatial research attempted to link the population surveys with the GIS and remote sensing data at discrete village locations and through the generation of continuous surfaces associated with village territories.

Spatial techniques were used to characterize village territories around nuclear villages by (1) creating overlapping territories by setting variable buffers as boundaries around village centers; (2) creating nonoverlapping village territories using Thiessen polygons for setting village boundaries; (3) growing village territories around village centers through fuzzy logic, in which transitional boundaries are defined; and (4) distributing the population at discrete village locations throughout village territories. Distribution was based on spread functions (e.g., inverse distance weighting) that use a mask or stratification to assign people to landscape strata based on land use and land cover types (e.g., to rice and cassava areas but not to water and forest sites) or other landscape weighting functions.

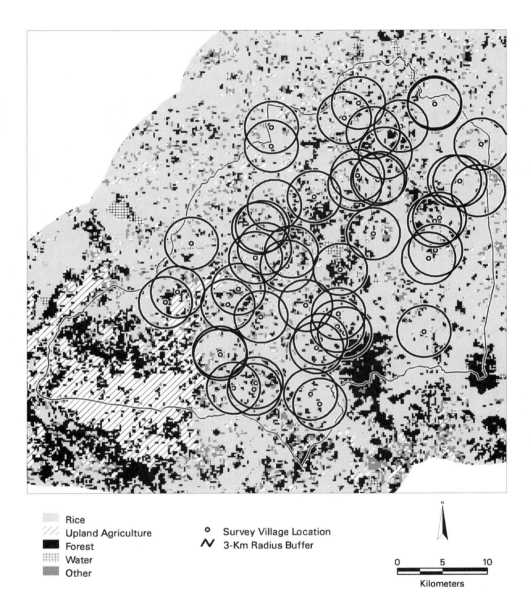

Rice
Upland Agriculture
Forest
Water
Other

° Survey Village Location
ᴎ 3-Km Radius Buffer

0   5   10
Kilometers

1993 LULC with 3km village buffers as territories.

Topographic Surfaces

The contour lines and point elevations on the 1984 1:50,000 scale base maps of the study area were digitally scanned, attributed, edge matched, and processed to yield DEMs of the study area. The construction of DEMs was important to characterize topography and relationships associated with land use potentials. Elevation, slope angle, and slope aspect were used to assess the local orientation of the landscape and generate slope convexity/concavity, soil moisture potential, and land use in defined village territories (e.g., percentages each of alluvial plain/lower terrace, middle/high terrace, and upland). Because of the physiographic relationships of terrain to moisture potential and soil suitability for either paddy rice or upland crops, terrain data were useful in landscape partitioning and for representing elements of the biophysical domain hypothesized to affect human behavior.

Topographic data inputs for DEM generation.

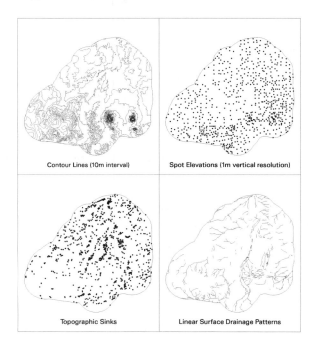

Topographic curvature integrates platform and profile curvature to portray localized topographic situation. Thus topography distinguishes slopes having a convex versus a concave slope. *Profile curvature,* the vertical or downslope concavity/convexity, portrays slope steepness. *Slope curvature* affects the acceleration and deceleration of gravity flows, including water movement across the surface. Soil moisture potential was estimated using an index of saturation potential and calculated using flow directions and accumulation grids based on eight directions for pixel-to-pixel flow routing. The Topographic Convergence or Wetness Index provides a simple and effective model for characterizing channels, gullies, and moisture sinks. The wetness index used calculations of slope and upslope contributing area (the area that drains through a given location) to generate a dimensionless index of potential wetness. The general form of the index follows:

$$Wi = ln\ (A/tan\ B)$$

where Wi = wetness index, ln = log normal, A = upslope contributing area, and B = surface slope.

Pattern Metrics

The concept of spatial metrics has emerged from landscape ecology. All landscapes are spatially organized, have functional interactions among the spatial elements that characterize them, and change over time. Pattern metrics quantify landscape structure, which is a prerequisite to studying landscape function and change. According to the operational paradigm, landscape form is indicative of its function, and land use and land cover patterns are the result of interacting processes of human, biophysical, and geographic dimensions. The metrics used in this study were run at the class (i.e., forest and agriculture) and landscape (i.e., village territory) levels within FRAGSTATS. Classified through satellite-time series data, land use and land cover were input to FRAGSTATS. The metrics were used to quantify the spatial and temporal changes in land use and land cover composition and pattern at the class and landscape levels as a signature of landscape form and function over time and space. Trajectories of change in the form of the landscape were related to changes in land function as defined through the population surveys. Site and situation measures of the land were defined through the GIS coverages of terrain, accessibility, and competition variables.

Geographic Accessibility and Competition

A measure of a village territory's accessibility to water and transportation was computed by defining the mean distance of each pixel associated with a particular village territory to, respectively, the nearest hydrographic feature and the nearest road within the network. Two land competition views were also computed: defining the number of village centers (point locations) that occurred within a prescribed village territory using a point-in-polygon approach, and defining the number of village territories that occurred within any portion of a prescribed buffer dimension around nuclear village centroids through a polygon overlay calculation. The accessibility measures define the ease of movement throughout the village territory, whereas village "competition" measures address the impact of village clustering on land cover conversion and the degree of competition for resources as a function of village density.

*Hydrographic accessibility
to district villages.*

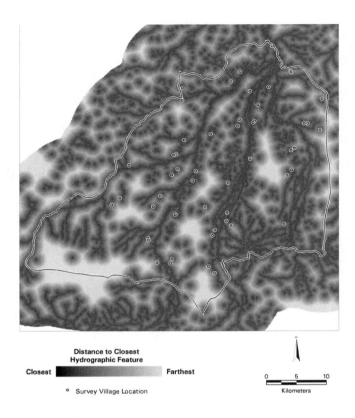

Distance to Closest
Hydrographic Feature

Closest ▬▬▬▬▬▬▬▬ Farthest

° Survey Village Location

0   5   10
Kilometers

Results
and Discussion

Population-environment interactions were and continue to be examined in the population, environment, and geographic domains using techniques such as population survey, remote sensing, GIS, and spatial and statistical analysis. This study outlines how the integrative capacity of GIS facilitates the organization of disparate data and how the spatial perspective was used to generate variables and test relationships.

Pattern metrics have been particularly useful in linking land use and land cover to population, environment, geographic accessibility, and competition. For example, the metric "percentage of landscape composed of cassava" indicates that the villages with more cassava production tended to have younger populations than those landscapes where more rice is grown. To some extent, these younger populations reflect the presence of young adults who otherwise might have moved away. As measured by percentage of land in forest in the mid-1970s, opportunities to expand cassava production had a negative effect on the out-migration of young persons between 1984 and 1994. Growth in the number of households in a village between 1984 and 1994 led to an increase in the percentage of village land devoted to upland crops by the latter date. "Cassava" villages tend to be larger than "rice" villages.

Landscape fragmentation measures showed that rice landscapes were much more homogeneous than cassava landscapes. Landscape diversity was found to be strongly correlated with the number of households in a village but weakly correlated with village population. These correlations with land cover classes were mirrored by the correlation between the number of patches in the landscape and the number of households pursuing rice cultivation. The number of patches in the landscape was highly negatively correlated with the mean distance to closest transportation feature. Landscape fragmentation was closely tied to road building. More remote villages tended to have more forest and less rice compared to less remote villages.

Adoption of a GIS framework for this study enriched the kinds of social and environmental interactions that could be addressed. Six obvious advantages of this approach include the ability to integrate disparate data; access to spatial tools and techniques; use of a relational database structure for linking population survey data to explicit spatial locations; the capability to represent temporal changes; the generation of variables for statistical modeling and sensitivity analyses; landscape representation using vector and raster data models; and presentation of results through maps, graphics, and statistical tables and equations.

The pressure placed on the land to yield greater agricultural output has resulted in a rapid expansion of land under cultivation, increases in agricultural inputs, and increased land fragmentation of ownership because of the shrinking availability of suitable land for agriculture. Migration to locate open land for agriculture as well as the deforestation of marginal sites, particularly in the middle and upper terraces, is a reaction to a marginal environment and as such leads to increased demand for subsistence agriculture, shrinkage of suitable land, and expanding market opportunities for agricultural products.

The linkage of population and environment within a GIS framework provides an opportunity to study variables in one domain through the lens of another domain. The overall intent of this research is to develop causal models of social behavior by integrating population-environment domains, simulating change in social systems through perturbations in the environment, and understanding the influence of internal and external forces on decisions affecting land use/land cover summarized at the patch, class, and landscape levels.

Acknowledgments

This essay is part of a larger set of interrelated projects funded by the National Institute of Child Health and Human Development (R01-HD33570 and R01-HD25482), the National Science Foundation (SBR 93-10366), the EVALUATION Project (USAID Contract #DPE-3060-C-00-1054), and the MacArthur Foundation (95-31567A-POP). The larger set of projects involves various collaborations among investigators at the University of North Carolina, Carolina Population Center, and the Department of Geography and investigators at the Institute for Population and Social Research (IPSR), Mahidol University, Bangkok, Thailand. Phil Page at the University of North Carolina provided general assistance with elements of this research.

References

Beven, K.J. and M.J. Kirkby. "A Physically-Based Variable Contributing Area Model of Basin Hydrology." *Hydrological Sciences Bulletin* 24:1 (1979), 43-69.

McGarigal, K. and B.J. Marks. "FRAGSTATS: Spatial Pattern Analysis for Quantifying Landscape." Version 1.0. Portland, Oregon: State University Forest Science Department, 1993.

Moore, I.D., R.B. Grayson, and A.R. Landson. "Digital Terrain Modeling: A Review of Hydrological, Geomorphological, and Biological Applications." *Hydrological Processes* 5 (1991), 3-30.

Parnwell, M.J.G. 1988. "Rural Poverty, Development and the Environment: The Case of North-East Thailand." *Journal of Biogeography* 15 (1988), 199-208.

Townsend, P.A. and S.J. Walsh. "Spatial Variability of a Wetness Model to Soil Parameter Estimation Approaches." *Earth Surface Processes and Landforms* 21:4 (1996), 307-26.

Societal Dimensions of Ecosystem Management in the South Florida Everglades

W.D. Solecki, Montclair State University, and R. Walker and S. Hodge, Florida State University

Resource Management Requirement

Understanding the pace and process of change affecting environmentally critical lands has become a priority in South Florida. General concepts associated with conservation strategies in the region (Yaffee et al. 1996, Christensen et al. 1996) have two principal goals: to understand and study the dynamic character of large-scale ecosystems, and to formulate plans that manage areas as an integrated whole instead of a patchwork of unassociated, small landscape elements.

GIS technology is a tool used by resource managers and research teams to examine ecosystem management issues. Uses include wetlands delineation, assessment of ecosystem characteristics, and biodiversity measurement. However, relatively little has been published regarding the role of GIS in examining the societal dimensions of ecosystem management, which include resource inventory and assessments, public outreach and participation programs, and analytical research. While the contribution of social science concepts and methodologies to ecosystem management has been articulated, the value of GIS in the process has not been fully explored. (See Machlis 1993 and Milon et al. 1997 for examples.)

This essay examines the interface between ecosystem management and social science research within a GIS framework. The discussion describes the way GIS methodologies were applied to the South Florida Everglades regional ecosystem as part of a "Man and the Biosphere" project focused on exploring ecological sustainability. (See Harwell et al. 1996 for a review of the project.) The usefulness of GIS in this broader project has two goals: providing a framework for characterizing natural resources and assessing societal impacts, and providing researchers and resource managers a platform for undertaking spatial analyses and regional ecosystem modeling.

Conceptualizing and Implementing Ecosystem Management

The concept of regional ecosystem management emerges from several decades of research on ecosystem form and function. It is based on the two principal findings listed below.

- Small-scale ecosystems function on an interdependent basis with adjacent small-scale ecosystems.

- Maintenance of small-scale ecosystems depends on sustainability of surrounding ecosystems, which together characterize a regional ecosystem.

Although the concepts of ecosystem management continue to evolve (see Slocombe 1993 and Grumbine 1994), several elements of its definition appear fixed. Ecosystem management is driven by explicit goals; executed by policies, protocols, and practices; and implemented based on ecological processes necessary to sustain structure and function (Christensen et al. 1996). Ecosystem management efforts incorporate the following imperatives.

- Design for long-term sustainability

- Establish clear operational goals

- Create sound ecological models that take into account complexity and interconnectedness

- Recognize the dynamic character of ecosystems

- Attend to context and scale

- Acknowledge humans as components of ecosystems

- Commit to management adaptability and accountability

Reaching these goals has been difficult in light of three sets of issues. One set focuses on conceptualizing ecological form and function in the face of inadequate knowledge regarding most systems. A data and information problem, this set of issues can only be resolved through field observation and data collection. A second set assesses existing management techniques with inadequate knowledge of system functions. The third set of issues, which relates to GIS, focuses on the appropriate spatial scale at which management practices should take place, and how human activity and occupancy should be regulated. Thus, ecosystem management demands extensive data collection and sophisticated analytical models. Managers and researchers must grapple with huge natural and social science data sets that have been prepared at different spatial scales and therefore have different accuracies. Contemporary landscape ecology is grounded in the notion that events at the local scale are interwoven with regional and global events and that ecosystems should be managed with multiple scales in mind. This observation brushes against the political reality that large-scale management initiatives are difficult to undertake given the probability of local opposition by a diversity of public interest groups.

Methodology

The main activities of the Human Dominated System workgroup in the South Florida Man and the Biosphere project have been directed at identifying the types and value of ecological sustainability data; determining whether such data were resident in the existing digital database; processing and evaluating available data; and making the data accessible to project participants. The GIS structure was designed to facilitate multifaceted, regionwide analyses focused on ecological and social processes and the interactions among them. Several types of data were identified for acquisition (see the next table). Primary classes of data included land use, vegetative cover, and human activity. Pursuant to these interests, land use data were assembled for two time periods to analyze temporal changes. Vegetative cover, or data that contained some vegetative description, provided information on gross cover type and change. Human activity data were primarily collected from U.S. Census records and included information on demographic characteristics, the economic base, and pertinent elements of the developed environment such as sewage disposal and sources of drinking water. Data were obtained at census block group, census tract, and county scales.

Types of data available and acquired through Man and the Biosphere project's Human Dominated System workgroup

Data	Description
Base map boundaries	Coverages serving as the base grid for the analysis. Although latitude and longitude grid data were available, the primary base grid consisted of 1980 census tract lines. Census tract lines for 1990 were collapsed into 1980 lines where possible. Tract centroids served as points from which to calculate distance.
Administration	Administrative boundaries within the biosphere reserve region. Includes boundaries for parks and other conservation lands, municipalities, and planning districts of federal, state, and local agencies.
Transportation	Primarily road network and highway systems. Sources identified through the U.S. Geological Survey (USGS), local and state governments, and other agencies.
Land use (1975)	Land use/land cover (LULC) data available through the USGS.
Land use (1986)	Land use data compiled by the South Florida Water Management District.
Land use (other)	A wide assortment of land use data available from other state and county planning agencies, particularly environmental management and planning divisions.
Ecology and physical geography topography	Vegetative land cover (types of vegetation found in the landscape) presumably constituting a continuous surface. Specific to the project, ecological endpoint data were also assembled. Topographic data provided elevations and contours important to hydrologic phenomena, vegetative cover, and other factors.

Data	Description
Hydrography	Locational information on bodies of water such as lakes and streams. Many data were available from the South Florida Water Management Division and USGS.
Demography and economics	Population data containing demographic, personal, and socioeconomic characteristics for geographic areas. Obtained from the U.S. Census Bureau, these data indicated the intensity of economic activity in manufacturing, agriculture, and service sectors.

Implementing the GIS

The first step was to design and inaugurate the GIS platform. U.S. Census Bureau TIGER (Topologically Integrated Geographic Encoding and Referencing)/Line files for the Everglades region were acquired. These included census tract files for the area spanning Broward, Collier, Dade, Hendry, Monroe, Lee, and Palm Beach counties. Spatial tract data were linked with nonspatial attribute data from the Census Bureau's Summary Tape File 3 (STF3) 1980 and 1990 data sets. Problems were noted and corrected. For example, from 1980 to 1990, some tracts were renumbered, partitioned, or reconfigured, and others split from one census to the next did not always adhere to uniform numbering conventions. To further complicate matters, in 1990 the Census Bureau extended tract lines offshore into coastal areas, a departure from earlier practices. (See the next illustration for 1980 and 1990 tract lines and histories.)

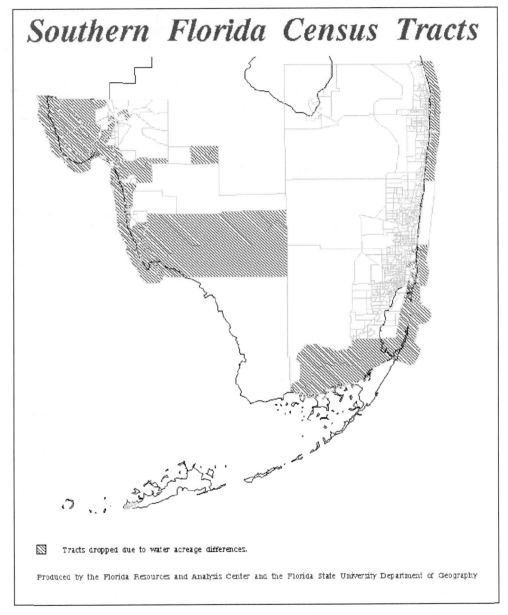

Southern Florida Census Tracts

▨   Tracts dropped due to water acreage differences.

(A)

Produced by the Florida Resources and Analysis Center and the Florida State University Department of Geography

South Florida census tract boundaries in 1980 and 1990 (A). Pathology tracts characterized by boundary changes and a greater than 5 percent difference in land use acreage (B). Pathology tracts with water acreage differences (C). Source: U.S. Census Bureau.

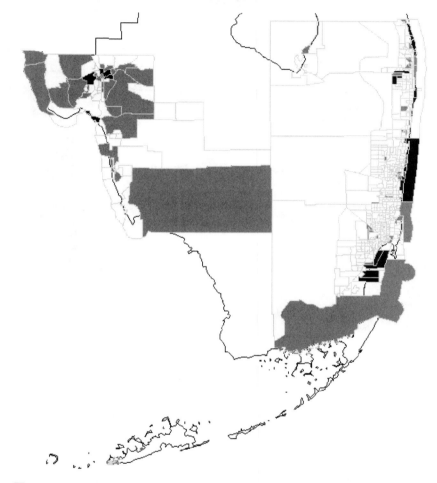

Southern Florida Census Tracts

■ Tracts dropped due to boundary changes

▦ Tracts with > 5% difference in landuse acreages.

(B)

Produced by the Florida Resources and Analysis Center and the Florida State University Department of Geography

Southern Florida Census Tracts

1990 Census Tract Lines

1980 Census Tract Lines

(C)

Produced by the Florida Resources and Analysis Center and the Florida State University Department of Geography

A major component of the GIS database was land use/land cover (LULC) change data. The USGS LULC 1:250,000 database (circa 1975) was overlaid with more recent land use data supplied by the South Florida Water Management District. Together they created a composite file consisting of more than 90,000 polygons.

Many other county and regional data sets from more than 20 federal, state, and local environmental agencies were collected and integrated into the GIS database. The data were divided into the three following categories.

- Base line: 1980 Census tract data served as the primary base for descriptive and analytical testing.

- Regionwide: Data available for the whole ecoregion, such as hydrology, soils, transportation networks, land use, and jurisdictional boundaries.

- Partial: Data available as higher resolution coverages for specific ecosystems, management areas, or land use categories.

The principal social science applications of the GIS concerned descriptive and inductive research and hypothesis testing.

Dependent and Independent Variables

A primary concern for ecosystem managers is determining causes of land use change. As a spatial modeling tool, GIS is one vehicle for answering this question. Regression analysis is a particularly useful part of the tool kit for understanding processes of land use conversion. In the Man and Biosphere project, an analysis was undertaken to examine the process of land use change through a regression model in which changes in native vegetation served as dependent variables, and a wide range of social forces and conditions served as independent variables.

Change from natural to human dominated land uses is believed to be a dependent variable driven by adjustments in social factors that affect the supply of and demand for land. Land *supply* is defined as a function of the areas available for development (including natural areas not under a form of public management), while land *demand* is a function of economic development potential. In South Florida, demand for land is determined by its potential for agricultural, commercial, industrial, or residential purposes. In a statistical model developed by Walker et al. (1997), a number of factors were considered possible determinants of demand.

Assumptions

The main independent variable was population growth. Certain aspects of community and human values were also assumed to play a role in conversion, including the educational level of residents, neighborhood stability, distance from core urban areas, and local zoning practices. Researchers also assumed the following hypotheses.

- Natural areas remain unchanged longer where stable, longstanding communities exist.

- Other factors remaining equal, individuals with more education exhibit a greater awareness for environmental impacts, which tends to reduce land conversion rates. (This effect is generally witnessed as citizen involvement in local land use planning decisions.)

- The rate at which undeveloped land changes is spatially linked to the extent of development of surrounding land and, in particular, distance from urban areas. Natural areas in the urban/agricultural fringe are likely to endure more intense conversion pressures than rural parcels given the accessibility of capital and proximity to urban centers.

In addition to traditional explanations and variables, a set of institutional factors linked to land development policy and environmental concerns also affect natural area conversions. For example, zoning regulations and the strength of environmental management organizations likely influence the supply of available land parcels and the demand shown by local citizens to develop them.

All Census data used to define independent variables were from the 1980 U.S. Census, except for population growth, which was constructed from 1980 and 1990 census data. Land use change (the dependent variable), generated by overlaying the 1975 and 1986 land use coverages at the census tract level, was defined as a switch from a natural cover type to any other human use—residential, agricultural, commercial, or industrial.

Results and Discussion

Maintenance of natural diversity and connectivity of regional landscapes has been defined as a necessary condition for ecosystem management in the Everglades. GIS technology was used in the Man and Biosphere project to identify how and why encroachment into the South Florida ecosystem occurred between 1975 and 1986. The 1975 and 1986 LULC data sets were classified into a set of common categories: natural drylands, natural wetlands, agricultural land, urban land, barren land, and water. These categories allowed for temporal analysis of land use/land cover change. By the mid-1970s, urban land uses were already established in the southeastern region, particularly around Miami, Fort Lauderdale, Naples, and Fort Myers. [See (A) in the next illustration.] Agriculture, widely present throughout the Atlantic coastal region, served as a transitional area between the highly urbanized shore and interior Everglades. With the exception of well-established agricultural areas south of Lake Okeechobee (and smaller and more dispersed agricultural areas just to the west), most of the remaining area was in a natural state. In addition, a patchy corridor still connected the large central contiguous wetland region with natural lands to the north.

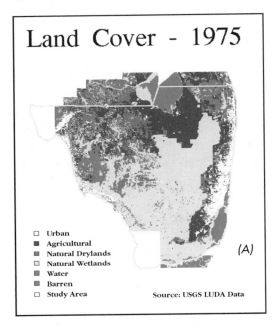

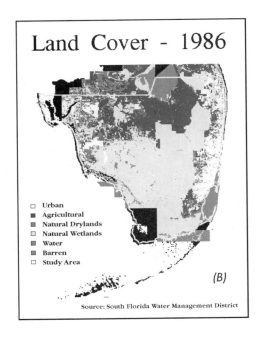

South Florida land use in 1975 (A) and 1986 (B).

By 1986, the basic pattern of land use remained the same. However, several important changes had taken place. [See (B) in the preceding illustration.] The agricultural zone separating Atlantic coast settlements from the interior Everglades was significantly reduced. More importantly with respect to ecosystem management, the connection between the Everglades and natural areas to the north was lost as natural lands were converted to agricultural and urban uses. In the west, urban and agricultural expansions created a highly fragmented landscape. Particularly in western interior areas, huge tracts of land were converted for agricultural production, leaving small and isolated pockets of undeveloped land.

Maintenance of natural wetlands and associated hydrocycles is especially critical for managing the regional ecosystem. GIS work was performed to document the amount and location of wetland fragmentation during the study period.

The 1975 and 1986 LULC data sets, which revealed a wide range of wetland categories, were recoded into a single category. Using the same approach as described above, an analysis of wetland change only was performed. At a regional scale, the physical integrity of Everglades wetlands (held mostly in public ownership) was maintained. However, wetland areas outside of the publicly held, central Everglades dramatically declined. (See the next image.) Most remnant wetlands within the eastern urban area disappeared between 1975 and 1986. In the northwestern region, wetland areas that helped to ecologically connect large natural areas were lost. The presence of these large, increasingly disconnected wetlands provided further evidence of ecosystem fragmentation.

Wetlands conversion in South Florida, 1975 to 1986.

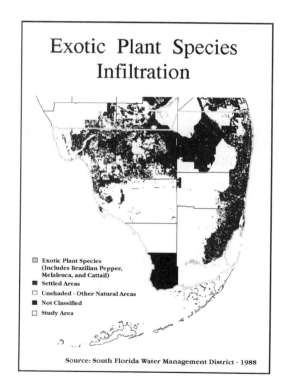

Exotic Plant Species Infiltration

☐ Exotic Plant Species (Includes Brazilian Pepper, Melaleuca, and Cattail)
■ Settled Areas
☐ Unshaded - Other Natural Areas
■ Not Classified
☐ Study Area

Source: South Florida Water Management District - 1988

Conclusions

The analytical results show that the rate of land use change was most positively associated with several variables, including the rate of immigration from outside South Florida and median household income. Outlying areas (affected by greater or lesser amounts of change) were also identified and are now the focus of follow-up research. In summary, the procedure developed for this study illustrates how results obtained through relational databases and mapping are linked to conceptual models and spatial analytical tests.

Acknowledgments

Funding for this study was provided, in part, by the U.S. Man and the Biosphere Program (Grant #1753-100110) to the University of Miami with a subcontract to Florida State University. U.S. MAB is administered by the U.S. Department of State as a multiagency, collaborative, interdisciplinary research activity to advance the scientific understanding of human/environment relations. This essay does not necessarily represent the policies of U.S. MAB, the U.S Department of State, or any member agency of U.S. MAB.

References

Christensen, N.L., A.M. Bartuska, J.H. Brown, S. Carpenter, C. D'Antonio, R. Francis, J.F. Franklin, J.A. MacMahon, R.F. Noss, D.J. Parsons, C.H. Peterson, M.G. Turner, and R.G. Woodmansee. "The Report of the Ecological Society of America Committee on the Scientific Basis for Ecosystem Management." *Ecological Applications* 6:3 (1996), 655-991.

Grumbine, R.E. "What is Ecosystem Management?" *Conservation Biology* 8:1 (1994), 27-38.

Harwell, M.A., J.F. Long, A.M. Bartuska, J.H. Gentile, C.C. Harwell, V. Myers, and J.C. Ogden. "Ecosystem Management to Achieve Ecological Sustainability: The Case of South Florida." *Environmental Management* 20:4 (1996), 497-521.

Machlis, G.E. "Social Science and Protected Area Management: The Principles of Partnership." *The George Wright Forum* 10:1 (1993), 9-19.

Milon, J.W., C.F. Kiker, and D.J. Lee. "Ecosystem Management and the Florida Everglades: The Role of Social Scientists." *Journal of Agricultural and Applied Economics* 29:1 (1997), 99-107.

Slocombe, D.S. "Implementing Ecosystem-based Management: Development of Theory, Practice and Research for Planning and Man" *Bioscience* 43:9 (1993), 612-23.

Walker, R.T., W.D. Solecki, and C. Harwell. "Land-Use Dynamics and Ecological Transition: The Case of South Florida." *Urban Ecosystems* 1:1 (1997), 37-47.

Yaffee, S.L., A.F. Phillips, I.C. Frentz, P.W. Hardy, S.M. Maleki, and B.E. Thorpe. *Ecosystem Management in the United States: An Assessment of Current Experience*. Washington, D.C.: Island Press, 1996.

People and Place:
Dasymetric Mapping Using ARC/INFO

S. Holloway, J. Schumacher, and R.L. Redmond, University of Montana

Resource Management Requirement

To better understand patterns of human settlement, migration, and related economic activities, social scientists traditionally have relied on data gathered directly from individuals and their families. Such census data can provide information about the socioeconomic characteristics of people living in different geographic areas at different times. In the United States, complete census data are collected every 10 years from people residing in geographic units defined as census blocks and delineated by the Census Bureau to contain approximately 100 people each. These enumeration units are nested hierarchically within block groups, census tracts, counties, and states. However, data analysis or display is difficult at the block level. Moreover, to ensure confidentiality, population is one of the few variables available for the smallest collection units. More detailed information about people and households is available only at higher levels of organization, starting with the block group.

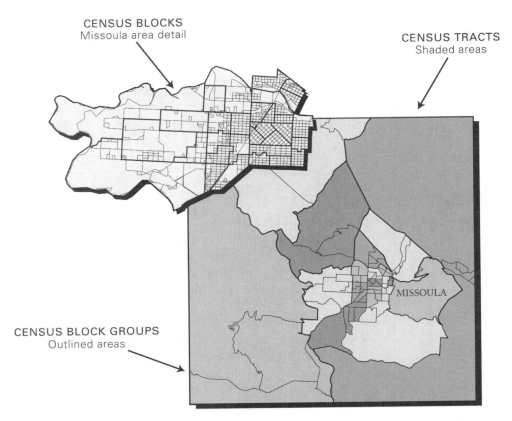

CENSUS BLOCKS
Missoula area detail

CENSUS TRACTS
Shaded areas

CENSUS BLOCK GROUPS
Outlined areas

MISSOULA

The hierarchical relationship among census blocks, block groups, and tracts is shown here for a portion of Missoula County, Montana. For the entire county, there are 74 block groups containing 2,238 blocks that fall within 18 census tracts.

Choropleth maps are the most common way to display census data, or any data for which the enumeration and mapping units are the same. For example, one can generate easily a choropleth map of human population density by dividing the number of people recorded in the enumeration unit (e.g., census block or block group) by the size of that unit. Despite their simplicity, choropleth maps have limited utility for detailed spatial analysis of socioeconomic data, especially in western North America, where human populations are concentrated in relatively few towns and cities found at lower elevations, and along major river corridors. Relatively large expanses of land are essentially uninhabited, especially in block groups or tracts distant from urban areas. When population density or any other socioeconomic variable is mapped by choropleth techniques, the results often reveal more about the size and shape of the enumeration units than about the people living and working within them.

One way to circumvent these limitations is to transform the enumeration units into smaller and more relevant map units in a process known as dasymetric mapping. This paper describes an automated GIS application that uses ARC/INFO to more precisely map human population density in Missoula County, Montana. The process is complex because population numbers from the 1990 census were reassigned to new map units using a combination of variables. Patterns of land ownership were used to identify uninhabited areas in each census block group. Land cover, land use, and topographic classes were used to further restrict the boundaries of inhabited areas. It is important to emphasize that other socioeconomic variables, such as median age of housing units, median income by ethnic group, or even births per 1,000 women, could be mapped according to the same basic approach wherever adequate data exist.

Methods

Population density data were derived from the 1990 U.S. Census, and the mapping unit was the census block group. Land ownership data were obtained as polygon coverages from the U.S. Forest Service, the U.S. Bureau of Land Management, and Plum Creek Timber Co.; source scale was 1:100,000 or finer. Topography was derived from U.S. Geological Survey (USGS) 7.5-minute digital elevation models (DEMs). Land cover was mapped from Landsat Thematic Mapper (TM) imagery using a two-stage classification procedure and a five-acre minimum mapping unit (MMU). Agricultural and urban land uses were manually labeled from the same Landsat TM imagery and at the same five-acre MMU.

These digital data were assembled into continuous (e.g., seamless) vector coverages in ARC/INFO version 7.0.4 running on an IBM RS/6000 UNIX workstation. Land cover, land use, land ownership, and DEM grids were intersected in the following sequence for each block group in the county (see the next illustration).

1. Areas uninhabited by people were identified by selecting census blocks with zero population; lands owned by the local, state or federal governments; corporate timber lands; and water features.

2. Four general land cover/land use classes were selected from the land cover data: urban, agricultural, forested, and open land.

3. All urban and agricultural polygons were assumed to be populated and left alone. Only areas of forested or open land with a slope less than or equal to 15 percent were "assigned" people.

4. Assuming that urban polygons contained more people per unit area than agricultural, wooded, or open polygons, the recorded population for each census block group was differentially allocated among these four types on a per-unit-area basis. A filtering routine was programmed into an AML to accomplish this task.

- Urban polygons were given a relative weighting of 80 percent (80 people per 100 population).

- Open polygons were weighted at 10 percent.

- Agricultural and forested polygons were weighted at 5 percent each. After each step, islands in the polygons resulting from each sequential intersecting operation were identified and flagged.

5. An attribute item for population density was added, and the database populated with the recalculated population density estimates.

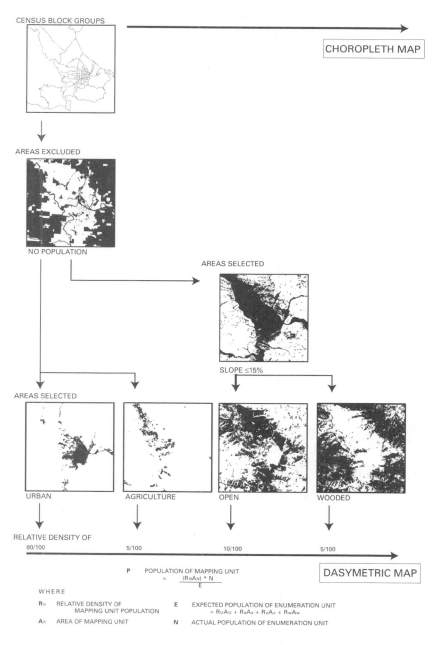

CENSUS BLOCK GROUPS

CHOROPLETH MAP

AREAS EXCLUDED

NO POPULATION

AREAS SELECTED

SLOPE ≤15%

AREAS SELECTED

URBAN   AGRICULTURE   OPEN   WOODED

RELATIVE DENSITY OF

80/100   5/100   10/100   5/100

DASYMETRIC MAP

P   POPULATION OF MAPPING UNIT
$$= \frac{(R_N A_N) * N}{E}$$

WHERE

R_N RELATIVE DENSITY OF   E   EXPECTED POPULATION OF ENUMERATION UNIT
         MAPPING UNIT POPULATION $= R_U A_U + R_A A_A + R_O A_O + R_W A_W$
A_N AREA OF MAPPING UNIT   N   ACTUAL POPULATION OF ENUMERATION UNIT

Filtering steps applied to each census block group to create a dasymetric map of population density.

Before an actual map could be made, the department selected appropriate density classes after examining a frequency histogram of all density values (see the next two figures). The number of classes, class breaks, and use of colors to indicate each class were all determined manually from the data distribution in the new mapping units. Finally, ARC/INFO polygons were exported as UNGENERATE files to a Macintosh platform and imported (using Avenza's Map-Publisher) into Adobe Photoshop and Adobe Illustrator to create the final maps.

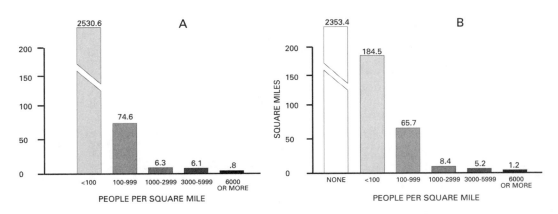

Area in Missoula County in each population density class derived from a choropleth map (A) and dasymetric map (B).

Results

The next image shows a choropleth map of population density for a portion of the county around the city of Missoula. Because of the way the enumeration units have been drawn, and because topography, land ownership, and land cover affect the distribution of people in the Missoula valley, the predominant class is the lowest one ("Less than 100"). In fact, this class represents nearly 97 percent of the entire county, covering 2,531 mi^2 (see A in the preceding illustration).

In contrast, the dasymetric population maps (see the next two images) identify a sixth class, with no population density ("None"), which covers 2,354 mi^2 or nearly 90 percent of the county. The "Less than 100" class represents only 185 mi^2, or 7 percent of Missoula County, which is a far more accurate picture of population density than the 97 percent indicated by the choropleth map. Changes in size for higher density classes appear relatively small at this scale, indicating that most of the differences between the two mapping techniques result from excluding unoccupied lands and, to a lesser extent, restricting slope. This finding alone indicates the importance of carefully excluding public and corporate lands where no people should be living.

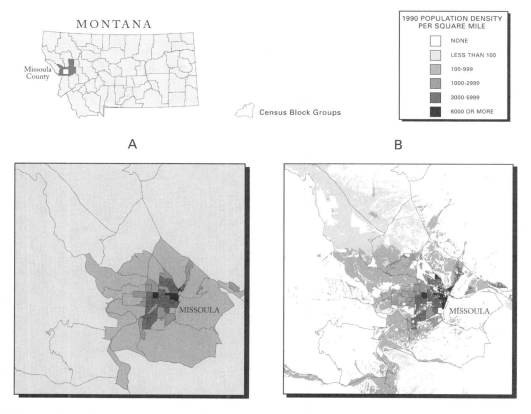

1990 population density for a portion of Missoula County in choropleth map (A) and dasymetric map (B).

Conclusions

In addition to population density, other socioeconomic variables such as income, gender, race, religion, occupation, and housing units also can be mapped using dasymetric techniques. This work is currently being extended to map 1990 population density, median age of housing units (by decade), and median income in a 45-county region in western Montana and northern Idaho. These results will enable land managers and policy makers to examine settlement patterns over time and in relation to distance from water, forest, or wilderness, and better understand why people live where they do in the western Rocky Mountains. Finally, when coupled with other broad-scale natural resource data that are becoming available in digital form, the database can serve as a predictive tool to identify areas most vulnerable to negative consequences of future development or land use change. In other words, when using these techniques, one is better able to ask and answer fundamental questions, such as, "Where is it?" "Why is it there?" and "How can we benefit from knowing that it's there?"

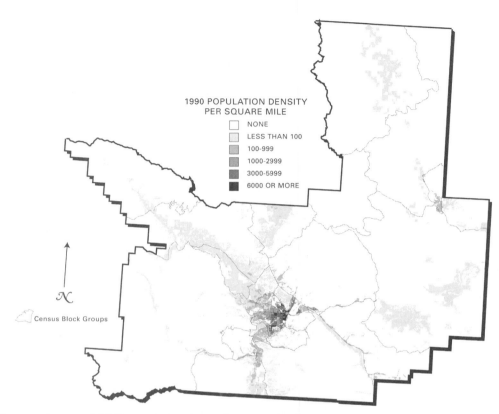

Dasymetric map of 1990 population density for Missoula County, Montana.

Acknowledgments This work was funded in part by the National Science Foundation (Grant # NSF ESR-9554501) and the U.S. Department of Agriculture, Forest Service (PNW-97-0511-2-RCRA). The authors benefitted from previous research on the topic by Yang Bin Tang, for which she received an M.A. degree in geography from The University of Montana. Finally, thanks to Jack Stanford (University of Montana, Flathead Lake Biological Station) and Cynthia Manning (Forest Service, Northern Region) for obligating funds to the project.

Modeling

W.W. Brady and G.L. Whysong, Arizona State University

The literature relevant to natural resources, GIS, and modeling is enormous even when restricted to the intersection of these three fields. Several factors have stimulated the accumulation of this literature (Jørgensen 1994). First, rapid growth in the power and availability of computer technology (hardware and software) is revolutionizing the ability to manage spatial data and simulate complex systems. At the same time, understanding of ecosystem complexity is increasing at a similar pace. Finally, the challenges facing resource managers have never been greater. These challenges range from traditional issues like sustained yield of timber to issues reflecting a new understanding of ecosystems such as protecting endangered species and habitats. Models always have been, and will continue to be, important tools in addressing the challenges natural resource managers face. Models provide tools that help scientists and managers to better understand the ecosystem as well as help forecast likely impacts of environmental change. The types of models available, however, are changing and the objective of the current chapter is to provide an introduction to this change.

Paradigm Shift

Natural resource management is in the midst of a paradigm shift (Franklin 1997). The old paradigm emphasized production of goods and services from forest and rangeland ecosystems, including timber, forage, water, wildlife, and recreation. The new paradigm emphasizes ecosystem management. While a universally accepted definition for ecosystem management remains elusive, the general parameters are clear. Management of ecosystems has become less anthropocentric and sustainable management of ecosystems (defined with reference to the ecological concepts of resilience and persistence) is now a major focus (Kaufmann et al. 1994, Pimm 1994).

This shift has resulted, in part, from a significant accumulation of knowledge over the last several decades concerning the structure and function of ecosystems. But the paradigm shift represents more than a simple reductionistic accumulation of knowledge about ecosystems. The fundamental perception of the ecosystem has changed. At present, the interdependence of ecosystems at numerous scales is better understood, ranging from the local (such as a meadow) up to the landscape, and ultimately to the biosphere itself. The omnipresence of change in ecosystems and the central roles feedback and disturbance play in the dynamics of natural systems is also better understood. The very concept of natural resources has expanded beyond commodities produced from ecosystems to include ecosystem processes (*natural services*) and biodiversity.

The new paradigm, furthermore, does not view human activity as exogenous to the ecosystem but as an integral element which must be understood to ensure ecosystem integrity. The inclusion of the human component further emphasizes the importance of scale in understanding ecosystem dynamics. Land use on the landscape is driven by economic, social, and institutional drivers which may reflect national or international policies and trends. For example, the relatively small, binational San Pedro watershed in southern Arizona and northern Sonora, has a *shadow* watershed which extends both to Washington D.C. and Mexico City.

Another feature of the new paradigm is a recognition of the importance of spatial relationships in ecosystem dynamics (Crow and Gustafson 1997). Landscape patterns and ecosystem processes are often interrelated in a complex manner dependent on the scale at which the landscape is considered. For instance, biodiversity is inherently a spatial concept that describes the abundance of species at multiple scales across a landscape. Biodiversity is also a function of disturbance which itself occurs at multiple spatial and temporal scales and intensities across a landscape.

The new paradigm both reflects and defines the issues which confront resource managers. While issues of production will continue to have a place in natural resource management, the overriding issues now are ecosystem protection and restoration. Franklin (1997) suggests that the basis for sustainability lies with two guiding principles: prevention of the degradation of the productive capacity of ecosystems, and prevention of the accelerated loss of genetic diversity. The issues which resource management must address are increasing both in terms of complexity and the breadth of temporal and spatial scale. Egler (1977) was prophetic when he remarked that not only are ecosystems more complex than we know, they seem to be more complex than we can know.

It is at this juncture that both modeling and GIS become important contributors to natural resource management. GIS contributes by providing a powerful tool for storage and analysis of spatial data. Models, on the other hand, facilitate organization of knowledge about ecosystems. Together they provide the tools for beginning to understand the complex temporal and spatial dynamics of ecosystems. Furthermore, models allow resource managers to ask questions about natural resources at spatial and temporal scales where extensive data collection is difficult, if not impossible.

Models

No universally accepted definition for the term *model* has emerged. Fowler (1997), for instance, defines a model as the representation of a process. Jeffers (1988) defines a model as the "formal expression of the essential elements of some problem in either physical or mathematical terms." Jørgensen (1994), on the other hand, describes a model as "a simplified picture of reality" used "as a tool to solve problems." Commonalities underlie all these (and other) definitions and center on the theme of goal-oriented representations of real world structures and/or processes. These representations have also taken numerous forms (from the physical to the mathematical). The models of concern here are spatially explicit mathematical models directed at natural resource management problems.

These models typically represent ecosystem structures and processes including the human elements and influences and share major components (Jørgensen 1994). *External variables,* or *forcing functions,* describe elements external to the model that influence the ecosystem. Typical examples of forcing functions would include precipitation or pollution inputs. Socioeconomic factors comprise another major set of external variables which many natural resource models must consider. *State variables* represent major structural elements of the system being modeled. *Mathematical expressions* are used to describe the relationships between external and state variables as well as interrelationships among state variables. These expressions take numerous forms from a simple graph to differential equations. Construction of mathematical expressions also requires identification of *parameters* which represent characteristics not expected to change with respect to the temporal or spatial dimensions of the model. For instance, water infiltration rate might be included as a parameter in a watershed model under the assumption of no change over the time scale and objectives of the model.

Good Models

Good models are characterized by a common set of features. At the most fundamental level models simplify the systems they are intended to represent. In the case of natural resources, the foundational ecological systems are far too complex for all structural or functional details to be included. In addition, the specific set of ecosystem elements selected for inclusion in a model is determined by the problem being addressed. For example, typical gap models used to model forest stand dynamics in response to disturbance (e.g., JABOWA in Botkin et al. 1972), FORET (Shugart and West 1977), or MIOMBO (Desanker and Prentice 1994) do not include the set of ecosystem elements required to model dynamics of resident bird populations in those same forest stands. Models, in other words, are not general representations of a system, and while they may share significant features, they are generally capable of addressing a rather limited, although perhaps very important, set of questions.

Simplification affects not only what ecosystem characteristics are included in a model, but also the temporal and spatial scales at which ecosystem structures and processes are represented. For example, general circulation models (GCMs), which attempt to represent the global climate system, often have a spatial resolution of $2°$ latitude by $2°$ longitude (Schneider 1993). A model detailing water resource dynamics in Colorado's Gunnison River Basin, on the other hand, requires a much finer spatial resolution (10km x 10km grids) (Hay et al. 1993). Neither model would be appropriate for understanding the movement of water through a particular soil type on a hillside. The same variations in temporal resolution exist. Models representing climatic processes which led to ice ages at intervals of about a hundred thousand years have very different temporal resolution than models concerned with next week's weather (Fowler 1997). Good models, therefore, are constructed to answer specific questions at specific spatial and temporal scales.

Good models can also be both verified and validated. Verification is a test of the internal logic of a model (Jørgensen 1994). Does the model give a reasonable representation of the structural and functional relationships of the real system? Relative to the model's purpose, have the proper forcing functions and state variables been included? Do the mathematical expressions reflect the best available scientific knowledge? Are estimates of model parameters reasonable? Are the spatial and temporal scales appropriate, given the modeling purpose? Verification, in this sense, is a logical evaluation of the model's assumptions. Good models would obviously reflect good science. Jeffers (1988) considers verification to be a subjective assessment of the model structure and behavior rather than an explicit comparison of model results against real world data.

Validation, on the other hand, is an objective test of model behavior (Jørgensen 1994). Not only should the assumptions of the model be logically and experimentally sound, but the results of the model, whether spatial or temporal predictions of ecosystem dynamics, should correspond in some sense to independent experimental data. However, validation is inherently difficult for several reasons. First, models are typically built to answer questions about the behavior of complex systems that are not fully understood. Therefore, data to allow a full validation of a model do not exist, or practically, cannot be collected. Second, it cannot be assumed that unusual behavior is indeed a symptom of model problems. Jeffers (1988) points out that numerous examples exist of models that display counterintuitive behavior. An important part of the validation process is making the distinction between unusual behaviors that reflect modeling difficulties and behaviors which reflect ecologically significant results. Collection of appropriate data for such a comparison requires careful attention to experimental design and is essential if model users are to have confidence in the results.

Finally, good models are transparent. Transparency guarantees that users of a natural resource model will know the specific problem a model was designed to address, assumptions made concerning relationships between ecosystem components and processes, and the temporal and spatial scales at which simulation occurs. If a model is not transparent, then users have no idea if the model is appropriate for their purposes. Opaque models raise the probability of falling into the classic GIGO (garbage in/garbage out) syndrome. The key to transparency is clear and readily accessible documentation and this should be a major goal of every modeling project. The importance of transparency can scarcely be overstated given the increasing importance of models in the decision making process (e.g., the importance of GCMs in understanding global climate).

It is important to emphasize that no one "correct" method exists for constructing models. No universally applicable set of rules exists and the "best" approach will depend on modeling objectives, existing data, and available modeling tools. The art of modeling is best learned by modeling. Beyond that, observation and critical evaluation of existing models are the best teachers (Fowler 1997). A wealth of models currently exists and is being rapidly augmented as the sophistication of computer hardware and software grows. Excellent resources for an introduction to models relevant to natural resource management and GIS include publications emanating from three conferences on GIS and environmental modeling (Goodchild et al. 1993, 1996). Our goal in the remainder of this chapter is to introduce readers to a representative set of existing models focusing on spatially explicit, dynamic models of particular importance to natural resource management.

Types of Models

Models relevant to natural resource management range from the relatively simple, such as the BIOCLIM model which has been used to determine the suitability of commercial tree species for transplantation (Booth 1991), to the extraordinarily complex general circulation models (GCMs) that attempt to represent the global climate system (Schneider 1993). Fowler (1997), Jørgensen (1994), Jeffers (1988), and Beltrami (1987) provide taxonomies for models. For our purposes, a simple classification would distinguish the descriptive versus explanatory, static versus dynamic, deterministic versus stochastic, simulation versus analytical, and spatially explicit models from those that are not.

Descriptive (or empirical) models only attempt to reproduce the behavior of a system with no underlying attempt to explain causality. For example, animal ecologists have long observed that population growth sometimes follows an S-shaped or logistic pattern (see the next illustration). A descriptive model of logistic growth can easily be created using the following polynomial equation:

$$N = b_o + b_1 t + b_2 t^2 + b_3 t^3 \qquad (1)$$

where N is population size and t represents time. This model will provide a good description of the observed pattern of growth but offers no insight into processes of population dynamics. Therefore, the applicability of this model to any other time, place, or population is absolutely unknown.

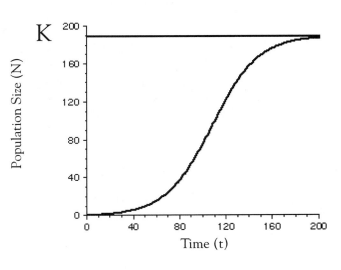

Logistic (S-shaped) growth curve illustrating a frequently observed pattern of population growth. K represents carrying capacity or the maximum number of organisms which can be supported in the habitat.

An *explanatory* model, on the other hand, attempts to explain a process by including the appropriate ecosystem structural and functional elements in the model. The following Verhulst-Pearl equation (written as a difference equation) is an example of a simple model which attempts to capture the basic population processes which lead to logistic growth:

$$N_{t+1} = N_t + rN_t(1-N_t/K) \qquad (2)$$

where N_t is population size at time t, r represents growth rate of the population, and K represents the carrying capacity of the habitat. The Verhulst-Pearl equation is extraordinarily simple; nevertheless, it captures important elements that influence the process of population growth. The strength of explanatory models lies in their extensibility, both in terms of complexity and generalization to new applications.

Static models represent a particular phenomenon at a point in time. These models may be viewed as a special case of simplification by the elimination of time. The appropriateness of a static model is entirely dependent on modeling objectives. If appropriate, static models have the advantage of being less data intensive and easier to develop compared to time dependent models. Most GIS programs and remote sensing packages allow users to build at least simple static models. In this case, the typical modeling objective is to build new GIS layers by overlay and algebraic manipulation of existing layers. Such models often have tremendous utility and are an important part of many modeling efforts. Morain and López-Baros (1996) describe numerous instances of using both vector GIS software and raster based image processing systems to construct static models. Gao et al. (1996) provide an excellent overview of GIS functionality for cell-based modeling. These applications range from business (e.g., choosing transportation routes, Nuñez Brown and Brinckerhoff 1996) to scientific (e.g., modeling vegetation, García and Murguía 1996). Jørgensen (1994) discusses both compartment model and response surface applications of static models. What this diverse collection of models has in common is the omission of temporal change.

Dynamic models, on the other hand, include temporal change as an important if not central component of the model (Hannon and Ruth 1997). For most natural resource applications, dynamic models are the appropriate choice because most problems involve change over time. The most common modeling approach is the use of differential or difference equations to describe changes in state variables. Because temporal change is important in natural resource issues, the emphasis below is on the principles of dynamic modeling and applications of these models.

Deterministic models contain no random variables. For instance, the Verhulst-Pearl equation discussed above is deterministic because given the same starting conditions, the results of the model will be identical every time. *Stochastic* models, on the other hand, do contain random variables. Therefore, the results of a stochastic model will not always be identical because random variables in the model potentially take on new values each time the model is run. The Verhulst-Pearl equation could be converted into a stochastic model by making carrying capacity (K) a function of precipitation (assuming precipitation is a random variable) as follows:

$$N_{t+1} = N_t + rN_t(1 - N_t/K_{pt}) \qquad (3)$$

where N_t is population size at time t, r is population growth rate, and K_{pt} is a random variable representing carrying capacity as a function of precipitation p at time t. Precipitation, in this case, would be an external variable to the Verhulst-Pearl equation and results of the model will not be identical every time except in the unlikely event that identical precipitation events occurred. The choice between deterministic and stochastic models is again a function of modeling objectives. Deterministic models are generally easier to build (Grant et al. 1997), particularly in terms of data requirements. However, to the extent that representation of the inherent variation in the real system is important, stochastic models are the better choice. Most natural resource models are stochastic because, as a general rule, extremes (expressed by the variation in a system) are more important in the dynamics of a system than average conditions. For instance, in a hydrologic model of a watershed, the extreme flows (highs and lows) are critical for understanding the dynamics of riparian vegetation communities.

Analytical models can be solved in closed form mathematically (Grant et al. 1997). Examples of analytic models are simple differential equations and response surface models. The Verhulst-Pearl model written in differential equation form follows:

$$\frac{dN}{dr} = rN\left(1 - \frac{N}{K}\right) \tag{4}$$

where the symbols have the same definitions as in equation (2). When the system being modeled can be represented by an analytical equation, then the solution to the modeling problem is greatly simplified. However, for most dynamic modeling situations, no general analytical solution exists and instead solution must be by numerical integration through computer simulation.

Simulation depends on numerical solution of models rather than analytic solution. Numerical solution of the Verhulst-Pearl equations is possible when they are written as difference equations rather than differential equations (equations 2 and 3). Simulation models, in effect, trace the behavior of the system through time, step by step. The disadvantage of simulation models has always been the lack of sufficient computer resources to implement anything approaching a realistic system. This disadvantage, however, is becoming less restrictive as available computer power continues to increase. The advantage of simulation models is the complexity which can be represented in the model.

From a natural resources perspective, *spatially explicit* models are of special interest. First of all, advances in ecosystem science have emphasized the importance of a spatial perspective. Second, remote sensing technologies, such as the Landsat and SPOT satellites have made immense amounts of spatial data available at reasonable cost and on a regular basis. Third, computer hardware and software are now available to manage and analyze these data. Many existing ecosystem models have concentrated on small-scale dynamics and have made the implicit assumption that spatial heterogeneity is not important (Gillman and Hails 1997). Beyond being an unrealistic assumption, it is also now an unnecessary assumption. One of the most exciting areas in natural resource management is the application of spatially explicit models to environmental and natural resource problems.

Dynamic Spatial Modeling

The previous sections have described GIS modeling using techniques that may be commonly available from scripting languages included within various GIS software packages. These techniques largely use algebraic methods to manipulate GIS layers for evaluation and predictive purposes.

A major limitation to such modeling approaches is that these models tend to be static or time invariant. That is, they are unable to take temporal change (spatial or otherwise) into account during a simulation exercise. For example, the temporal change in urban land use (U) is not only influenced by physical orographic features, such as topography (T), but also land ownership (O), current land use (Lu), available transportation (Tr), human population dynamics (P), plus social (S) and economic (E) factors. These factors interact through processes that influence changes in spatial land use involving both deterministic and stochastic mechanisms. As the process of spatial urbanization proceeds through time (t), the probability that a specific undeveloped area will be urbanized at any point in time depends on the factors and processes listed above, many of which are also changing through time. Dynamic models are capable of taking such changes into account, since they describe the process of change. For example, modeling the urban land use process involves mathematically describing the rate of urbanization with respect to time. This may be illustrated in a generalized fashion as follows:

$$\frac{\Delta U}{\Delta r} = f(P, O, Lu, Tr, T, S, E) \qquad (5)$$

and

$$U_t = \frac{\Delta U}{\Delta t} + U_{t-1} \qquad (6)$$

where symbols are as described above. The solution of the model involves numerical integration of the above and supporting equations through time and space. The spatial component can be provided through the use of GIS layers that document the input variables, which, in many cases, are themselves as dynamic as the variables presented above. Consequently, the solution of such models in space and time is best suited to raster based GIS layers where model equations are applied to each location (cell) of the appropriate layer. Thus, such models may be termed "cell models." Clearly, the brief spatial modeling example presented above can be expanded to consider virtually any mix of natural resources in a spatial context where appropriate knowledge and spatial information are available, or can be acquired from existing data or through remote sensing techniques.

The simulation of dynamic spatial models commences with the current state (condition) of a landscape. Typically, this state is characterized by raster based GIS layers describing important landscape components (land use, vegetation, topography, etc.). The formulation and simulation of such models through time can provide information about how natural processes can be influenced and allow evaluation of plans (scenarios) for future landscape status and use. The entire process assists in identifying the relative importance of different types of data (information) to system dynamics through sensitivity analysis. The contributions of individuals from different disciplines, combined with formalized concepts of how the landscape works, often result in increased knowledge and appreciation of how these complex systems function. Although spatial dynamic models are capable of taking dynamic change into account during a simulation, they do not tend to optimize planning, predict the future with great accuracy, or tell a land manager the solution to a specific problem. However, simulation exercises are valuable for providing information concerning future trends. This may improve the information base for land managers, assist in making wise decisions, and help avoid potentially undesirable consequences.

Dynamic Models

The ecosystem concept as originated by Tansley (1935) has become instrumental to the transition of ecology from a descriptive science to a predictive science capable of answering questions involving ecosystem dynamics and change (Shugart 1998). Ecosystem models of plant and animal populations have played an important role in ecology. A number of ecologists have developed independent models of animal population dynamics (Huston et al. 1988, Holling 1961 and 1964, Rohlf and Davenport 1969). Some of the earlier work on plant species involved growth of individual trees in forest based models (Newnham 1964, Lee 1967, Mitchell 1969, Lin 1970, Bella 1971, Arney 1974, Hatch 1971).

Early work incorporating a spatial component into landscape models involved Markov transitional probabilities and differential or difference equations as applied to various classifications of landscape type (mosaic, interactive, or homogeneous). An excellent reference to this and more recent work is provided by Shugart (1998). The availability of low-cost computers and the development of GIS software have allowed advances in how spatial data may be acquired, displayed, and analyzed. Although GIS/remote sensing software can display and analyze time series spatial information, it does not provide suitable dynamic modeling tools. Ecologists, however, were quick to recognize the potential for integrating their dynamic landscape modeling efforts with GIS. Early approaches to incorporate this technology were developed and applied in several parks in the United States and Australia (Cattelino et al. 1979, Kessell 1979).

Unfortunately, efforts to combine temporal dynamic models with a spatial component proved to be a laborious task because the tools required to integrate model software with GIS software were unavailable. Consequently, in most cases, significant programming and computer skills were required to integrate computer software. The interpreted scripting and macro languages available within existing GIS software proved too slow for serious spatial modeling. A comparison of cpu resources required to simulate acid deposition of sulphur dioxide using a spatial 70 x 80 raster grid clearly indicated the superior speed obtained by coupled GIS and compiled languages (Dragosits et al. 1996). In addition, lack of spatiotemporal data representation within current GIS software is a major disadvantage for dynamic modeling purposes (Yuan 1996).

During recent years, significant progress has been made in developing software that allows integration of GIS layers with model development and simulation. The Land Use Change Analysis System (LUCAS) interfaces with the public domain GRASS GIS-plus-database software for evaluating land use change by considering socioeconomic factors and transitional probability matrices, along with spatial land classification criteria that influence land use decision making (Berry et al. 1996). Amazonia is a computational modeling system designed for modeling large-scale earth science systems (Smith et al. 1996). The spatial modeling environment (SME) integrates GRASS and a generic object database into a complete modeling environment that can also operate in parallel computing environments (Costanza and Maxwell 1991, Maxwell and Costanza 1994, 1995 and 1996). Model construction for the SME is facilitated by using the systems dynamics approach (Forrester 1961) and a specialized modeling language, DYNAMO (Richardson 1981). Currently, versions of DYNAMO that are extremely easy to learn and use are available for Windows and MacIntosh computer systems such as STELLA (HPS 1995). Both LUCAS and SME can be implemented in parallel computing environments using parallel virtual machines (PVM) and message passing interface (MPI) software in UNIX environments, respectively. This can be an important consideration for practical simulation of medium- to large-scale dynamic spatial models. A spatially explicit landscape event simulator (SELES) has been developed to evaluate event driven disturbance using either deterministic or stochastic processes on a landscape scale (Fall and Fall 1996).

Efforts have also progressed to integrate existing models into an intercommunication framework of which the Forest Ecosystem Dynamics Modeling Environment (FEDMOD) is an example (Knox et al. 1996). Westervelt and Hopkins (1996) discuss linking multiple landscape simulation software packages into an integrated spatiotemporal ecological modeling system (I-STEMS) for purposes of simulating the behavior of individual animals. This effort builds on the use of GRASS and SME. A spatially explicit version of the CENTURY ecosystem model can be used to simulate ecosystem dynamics within regions (McKeown 1996). One limitation of this model is that it considers raster cells as independent during model simulation; thus, movement of material across a landscape is not easily considered. A dynamic spatial modeling capability has recently been added to PCRaster (Wesseling et al. 1996). This software runs on DOS and UNIX systems and incorporates spatial statistics as well.

Anatomy of a Dynamic Spatial Model

Any modeling project must start with the definition of specific modeling objectives and the important processes involved. Only components and variables important to explaining processes should be included. Generally, the inclusion of variables with limited influence on a particular process dramatically increases model complexity but provides little to no real change in model behavior. Variables and data that provide the greatest influence on a process must be included for the model to be realistic and accurate. In addition, clear boundaries as to what will and will not be considered in the model must be identified.

Individual processes can usually be subdivided into submodels that pass data and information to each other during a simulation. As an example, consider the urban land use submodel shown in the next figure. The systems dynamics approach is used for illustrative purposes, although in a simplified form. Several examples of the systems dynamics approach to ecology and natural resource management are presented by Grant et al. (1997). The figure also illustrates the flowcharting of how a process or submodel is envisioned to work and the communication between submodels that may comprise the entire model. The flowchart can be considered a blueprint for model construction. The urban land use submodel is simply a portion of a larger model that encompasses human population dynamics, social and economic factors, and air and water pollution, in addition to land use. The state variables in the model are considered physical measurable entities that may change dynamically in a temporal fashion. Only a fixed spatial area, as defined by the modelers, is considered. Such area could be limited to urban boundaries or encompass entire watersheds. The land area can be in a variety of land use states (uses) at any given time, of which "urban" use may be only one. Even the urban land use state variable could be further divided into state variables describing commercial and/or residential use. If the only purpose of the model is to describe the process of urbanization, however, such detail may be undesirable because it would result in a corresponding increase in overall model complexity and data requirements. The time and cost of constructing models characterized by high complexity and data requirements would require serious evaluation of practicality and feasibility.

The process being modeled in the urban land use submodel is the rate of urbanization. The purpose of other state variables and submodels is to provide information necessary to mathematically describe this process or rate. Such a procedure forces one to attempt to learn and understand the important factors involved and how the process of urbanization actually works. The variable "time" should never be used as an independent variable in these relationships, since the resultant mathematical equations will tend to decompose to functions of time and fail to describe the real process being modeled or increase understanding of the process. At least one potentially serious flaw exists in the urban land use submodel illustrated in the following figure. Once land use is characterized as urban (urban land use state variable), the model does not consider that urban land could be used for anything other than urban purposes. Thus, the model considers urbanization as a unidirectional process only. While this may be true for rapidly expanding urban environments, it may be questionable for long-term ecological considerations.

Simplified flowchart illustrating possible relationships among components influencing the urban land use process using the systems dynamics approach.

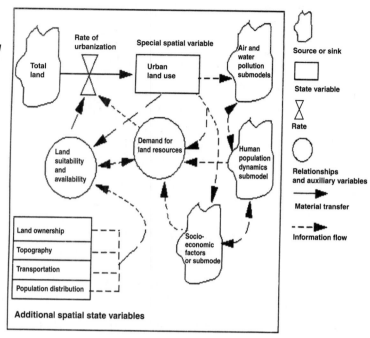

The state variables in the previous illustration are termed spatial state variables. Thus, they comprise data concerning attributes of the total landscape being considered by the model. Each is a GIS raster layer which will provide the initial (starting) conditions for the model at the same specific point in time (a snapshot). Although both raster and vector based GIS software exist, raster based GIS provides a good environment for spatial ecological modeling, since it includes numerous spatial functions useful for dynamic modeling (Slothower et al. 1996). Time series spatial data can often be of importance, because the final spatial model can be simulated using the earliest available data as a starting condition and model output compared to actual historical information. Such a procedure can be extremely useful for model adjustment. However, any model can be forced to fit the data used to build it; consequently, such a procedure cannot be considered model validation. For true validation, a completely independent data set is required.

Actual construction, simulation, and preliminary evaluation of the model can commence prior to incorporating spatial data. However, the type and range of spatial information expected will be required as inputs for simulation and evaluation. Consequently, initial model development using dynamic modeling software such as STELLA (HPS 1995) or Vensim (VSI 1998) may be conducted in parallel with GIS preparation of raster layers for the final spatial model.

Dynamic spatial models require large amounts of data. Numerous raster GIS layers may be required to provide adequate information for model equations and parameter estimation. Improvements in type and availability of spatial information from local, state, and federal sources, plus the decreasing cost of remote sensed information suitable for GIS classification can help, but publicly available sources are unlikely to meet all data requirements.

A representation of the dynamic spatial model simulation of the urban land use submodel is illustrated in the next figure. The figure is an oversimplification of the spatial simulation process, since it does not consider the complete model that would actually be simulated simultaneously. However, sufficient information is presented to illustrate the simulation process.

Diagrammatic representation of dynamic spatial simulation of an urban land use submodel illustrating GIS and database inputs, cell model solution, GIS output, and temporal update to the model.

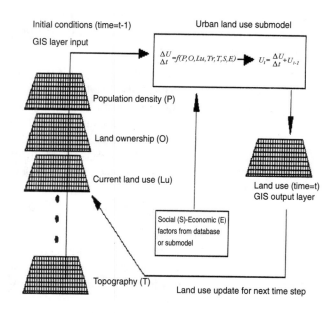

Initial conditions (time=t-1)

GIS layer input

Urban land use submodel

$$\frac{\Delta U}{\Delta t} = f(P, O, Lu, Tr, T, S, E) \rightarrow U_t = \frac{\Delta U}{\Delta t} + U_{t-1}$$

Population density (P)

Land ownership (O)

Current land use (Lu)

Land use (time=t)
GIS output layer

Social (S)-Economic (E) factors from database or submodel

Topography (T)

Land use update for next time step

For purposes of discussion, it is assumed that the relevant raster GIS layers for the initial model conditions, any relevant database information, and the model itself have been prepared. During simulation, all these data sources must be available. The urban land use submodel equations require information from each cell of each relevant GIS layer, as well as from other linked submodels or a database. Each raster cell over the landscape is processed during a single time step (from time *t-1* to time *t*). For example, raster layers describing land ownership and current land use may suggest that a particular location (raster cell) is suitable to meet the demand for increased urban growth. However, another GIS layer may exclude the cell due to topography (slope) or because the cell is part of a flood plain. In addition, the state of adjacent cells can be considered. For example, a location in close proximity to an already developed area is more likely to become urbanized simply due to availability of services. Once the model finishes a time step, the land use layer is updated to reflect the change in urban land use. The model may continue into the next time step and/or the new land use layer can be imported into a GIS for analysis purposes.

Individual integration of raster GIS and database information to the spatial simulation model is not an easy task and requires suitable computer programming skills. A much simpler alternative is to use a dynamic, spatial modeling environment, such as SME, which automatically integrates the GIS, database, and model components while controlling the entire simulation process.

Computational Considerations

One of the major problems associated with incorporating dynamic models into a GIS is that current GIS software is not designed to numerically integrate the complex mathematical interrelationships among GIS layers and external driving forces. Although most GIS software packages include scripting or "modeling" languages for writing user macros, they are ill suited for even small-scale dynamic spatial simulations. One reason is that these languages tend to be interpreted rather than compiled. Programs written in interpreted languages execute much slower than compiled programs. This is particularly important for computationally intensive programs. All dynamic modeling is computationally intensive. When the spatial component is added, computer resources required for a timely solution increase dramatically. A second reason is that some GIS software is available for a limited variety of computer operating systems. Some very popular operating systems are not well suited for interprocess communications, networking, and multi-tasking—all of which are needed for serious dynamic spatial modeling. This is exemplified by considering that thousands of computer calculations may be necessary for the solution of a model at one location (cell) to move one step in the temporal dimension. A small spatial grid of 100 x 100 cells would increase the computational load by a factor of 10,000 (without taking input-output overhead into consideration). Consequently, a dynamic model, written in a compiled language and having no spatial component, that takes 5 seconds to simulate a 10-year time period could be estimated to take in excess of 13 hours for a 100 x 100 grid. Most spatial dynamic modeling efforts will involve a larger spatial representation. Dragosits et al. (1996) indicated that an ARC/INFO AML spatial simulation of a 70 x 80 cell grid used about 500 times more computer resources as an identical coupled FORTRAN implementation.

The spatial resolution used for model simulation can have a major impact on computer resources. If a 100 x 100m raster grid will suffice, the selection of a 30 x 30m resolution will require dramatic increases in computer resources. Selection of cell resolution for dynamic spatial modeling is an important issue. First, all GIS raster layers should have the same resolution. Because each cell defines a specific location of known area, model solution is cell specific. Second, there is almost nothing to be gained by using a higher resolution than provided by the best distribution of data across a GIS layer. A large loss in computer performance is guaranteed if higher spatial resolutions are chosen for "aesthetic" reasons. Furthermore, large raster data sets require more disk space, memory, and cpu resources for storage and processing, resulting in the need for more expensive and powerful computing platforms.

Available computer resources are an important consideration for any dynamic spatial modeling endeavor. Berry et al. (1996) provide a brief comparison of simulation times between a single SPARCstation 5 as compared to a cluster of 16 machines running a LUCAS simulation and clearly illustrate the advantage of using multiple machines in parallel to solve dynamic spatial models. The computer industry is increasingly making available larger, more powerful computers at lower prices. Personal computers can match or exceed the capabilities of some dedicated computer workstations at lower cost. The powerful UNIX environment, within which much serious dynamic spatial modeling is done, is now available for PCs. FreeBSD and Linux are available for PCs at virtually no cost. The spatial modeling environment (SME) software (among others) is available free of charge for noncommercial purposes (educational, research, and nonprofit) and can be compiled on a wide variety of UNIX operating systems, including Linux. PVM and MPI are freely available and can be used for communications between multiple machines running a spatial simulation. Cooperating individuals can use their PCs to construct a network of workstations (NOW) to significantly improve simulation speed while maintaining computer resources for daily usage. Beowulf class clusters of PCs can be constructed to achieve supercomputer capabilities for a fraction of traditional supercomputer cost (Ridge 1997). In fact, a bit of creative effort resulted in a Beowulf cluster of PCs at Oak Ridge National Laboratory that currently consists of 126 nodes at literally no cost for computer hardware or software (ORNL 1998). Recently, NASA joined with Red Hat Software Inc. to distribute on CD the complete software and documentation from their pioneering Beowulf project (Red Hat 1998). The CD includes the Linux operating system adopted by the NASA Goddard Space Flight Center for the Beowulf project. Clearly, low-cost supercomputer power and software are available for dynamic spatial modeling purposes, if one is willing to put some time and effort into assembling the computer resources. For the novice, a bit of luck would be helpful as well.

WWW Resources

The Internet contains a great deal of information relative to dynamic modeling, systems dynamics, dynamic spatial modeling and the computer resources previously discussed. Selected sources are presented below for those who wish to pursue these topics further.

Easy to learn and use systems dynamics software (commercial) that can be used with the spatial modeling environment (SME) include STELLA (*http://www.hps-inc.com/*) and Vensim (*http://www.std.com/vensim/*).

Currently, the Vensim Personal Learning Edition (PLE) is a freely available version for download. Web pages for several spatial modeling software packages described earlier are LUCAS (*http://www.cs.utk.edu/~lucas/*), Century (*http://www.nrel.colostate.edu/PROGRAMS/MODELING/CENTURY/CENTURY.html*), PCRaster (*http://www.frw.ruu.nl/pcraster/pcraster.html*), and SME (*kabir.cbl.umces.edu/SME3/index.html*).

Information about ORNL's supercomputer may be found at *http://www.esd.ornl.gov/facilities/beowulf/*. Beowulf supercomputer project information may be accessed at *http://cesdis.gsfc.nasa.gov/linux/beowulf/beowulf.html*. Information on the Extreme Linux software (CD) for integrating multiple PCs into a NOW or Beowulf class supercomputer may be found at Red Hat Software (*http://www.redhat.com/*).

Several references cited in this chapter are from the National Center for Geographic Information and Analysis (NCGIA) Third International Conference/Workshop on Integrating GIS and Environmental Modeling. The proceedings are published in electronic form (*http://bbq.ncgia.ucsb.edu/conf/SANTA_FE_CD-ROM/main.html*) and are also available on CD from the NCGIA (*ncgia@ncgia.ucsb.edu*).

References

Arney, J.D. "An individual tree model for stand simulation in Douglas-Fir." In J. Fries, ed., *Growth models for tree and stand simulation research notes,* pp. 38-46. Stockholm, Sweden: Department of Forest Yield Research, Royal College of Forestry, 1974.

Bella, I.E. "A new competition model for individual trees." *Forest Science* 17 (1971), 364-72.

Beltrami, E. *Mathematics for Dynamic Modeling.* Boston: Academic Press, 1987.

Berry, M.W., R.O. Flamm, B.C. Hazen and R.L. MacIntyre. "Lucas: A system for modeling land use change." *IEEE Computational Science & Engineering* 3:1 (1996), 24-35.

Booth, T.H. "A climatic/edaphic database and plant index prediction system for Africa." *Ecological Modelling* 56 (1991), 127-34.

Botkin, D.B., J.F. Janak and J.R. Wallis. "Some ecological consequences of a computer model of forest growth." *Journal of Ecology* 60 (1972), 849-72.

Cattelino, P.J., I.R. Noble, R.O. Slatyer and S.R. Kessell. "Predicting the multiple pathways of plant succession." *Environmental Management* 3 (1979), 41-50.

Costanza, R. and T. Maxwell. "Spatial ecosystem modeling using parallel processors." *Ecological Modelling* 58 (1991), 159-83.

Crow, T.R. and E.J. Gustafson. "Concepts and Methods of Ecosystem Management: Lessons from Landscape Ecology." In M.S. Boyce and A. Haney, eds., *Ecosystem Management: Applications for Sustainable Forest and Wildlife Resources,* pp. 54-67. New Haven: Yale University Press, 1997.

Desanker, P.V. and I.C. Prentice. "Miombo—A Vegetation Dynamics Model for the Miombo Woodlands of Zambezian Africa." *Forest Ecology and Management* 69 (1994), 87-95.

Dragosits, U., C.J. Place and R.I. Smith. "The potential of GIS and coupled GIS/conventional systems to model acid deposition of sulphur dioxide." NCGIA. Proceedings Third International Conference/Workshop on Integrating GIS and Environmental Modeling. January 21-25, 1996, Santa Fe, New Mexico.

Egler, F. *The Nature of Vegetation: Its Management and Mismanagement.* Norfolk, Connecticut: Aton Forest, 1977.

Fall, J. and A. Fall. "SALES: A spatially explicit landscape event simulator." NCGIA. Proceedings Third International Conference/Workshop on Integrating GIS and Environmental Modeling. January 21-25, 1996, Santa Fe, New Mexico.

Forrester, J.W. *Industrial Dynamics.* Cambridge, Massachusetts: MIT Press, 1961.

Fowler, A.C. *Mathematical Models in the Applied Sciences.* Cambridge, Massachusetts: Cambridge University Press, 1997.

Franklin, J.F. "Ecosystem Management: An Overview." In M.S. Boyce and A. Haney, eds., *Ecosystem Management: Applications for Sustainable Forest and Wildlife Resources,* pp. 21-53. New Haven: Yale University Press, 1997.

Gao, P., C. Zhan, and S. Menon. "An Overview of Cell-Based Modeling with GIS." In M.F. Goodchild, L.T. Steyaert, B.O. Parks, C. Johnston, D. Maidment, M. Crane, and S. Glendinning, eds., *GIS and Environmental Modeling: Progress and Research Issues,* pp. 325-31. Fort Collins, Colorado: GIS World Books, 1996.

García, D., F. Lozano and R.G. González Murguía. "Modeling Vegetation Distribution in Mountainous Terrain." In S. Morain and S. López Baros, eds., *Raster Imagery in Geographic Information Systems,* pp. 165-71. Santa Fe, New Mexico: OnWord Press, 1996.

Gillman, M. and R. Hails. *An Introduction to Ecological Modelling.* Oxford: Blackwell Science, 1997.

Goodchild, M.F., B.O. Parks, and L.T. Steyaert, eds. *Environmental Modeling with GIS.* New York: Oxford University Press, 1993.

Goodchild, M.F., L.T. Steyaert, B.O. Parks, C. Johnston, D. Maidment, M. Crane, and S. Glendinning, eds. *GIS and Environmental Modeling: Progress and Research Issues.* Fort Collins, Colorado: GIS World Books, 1996.

Grant, W.E., E.K. Pedersen and S.L. Marín. *Ecology and Natural Resource Management: Systems Analysis and Simulation.* New York: John Wiley & Sons, 1997.

Hannon, B. and M. Ruth. *Modeling Dynamic Biological Systems.* New York: Springer-Verlag, 1997.

Hatch, C.R. "Simulation of an Even-aged Red Pine Stand in Northern Minnesota." Ph.D. diss., University of Minnesota, 1971.

Hay, L.E., W.A. Battaglin, R.S. Parker and G.H. Leavesley. "Modeling the Effects of Climate Change on Water Resources in the Gunnison River Basin, Colorado." In M.F. Goodchild, B.O. Parks and L.T. Steyaert, eds., *Environmental Modeling with GIS*, pp. 173-81. New York: Oxford University Press, 1993.

Holling, C.S. "Principles of Insect Predation." *Annual Review of Entomology* 6 (1961), 163-82.

_____. "The analysis of complex population processes." *Canadian Entomology* 96 (1964), 335-47.

HPS. "STELLA: High Performance Systems." URL: http://www.hps-inc.com/, 1995.

Huston, M., D.L. DeAngelis and W.M. Post. "New computer models unify ecological theory." *Bio Science* 38 (1988), 682-91.

Jeffers, J.N.R. *Practitioner's Handbook on the Modelling of Dynamic Change in Ecosystems*. New York: John Wiley & Sons, 1988.

Jørgensen, S.E. *Fundamentals of Ecological Modelling,* 2nd ed. Developments in Environmental Modelling 19. Amsterdam, The Netherlands: Elsevier Science, 1994.

Kaufmann, M.R., R.T. Graham, D.A. Boyce, Jr., W.H. Moir, L. Perry, R.T. Reynolds, R.L. Bassett, P. Mehlhop, C.B. Edmister, W.M. Block, and P.S. Corn. "An ecological basis for ecosystem management." U.S. Forest Service General Technical Report RM-246, 1994.

Kessell, S.R. *Gradient Modeling: Resource and Fire Management*. New York: Springer-Verlag, 1979.

Knox, R.G., V. Kalb, E.R. Levine and U. Bindingnavle. "A framework for integrating environmental models to simulate forest ecosystem dynamics." NCGIA. Proceedings Third International Conference/Workshop on Integrating GIS and Environmental Modeling. January 21-25, 1996. Santa Fe, New Mexico.

Lee, Y. "Stand models for lodgepole pine and limits to their application." *Forest Chronicles* 43 (1967), 387-88.

Lin, J.Y. "Growing Space Index and Stand Simulation of Young Western Hemlock in Oregon." Ph.D. diss., Duke University, 1970.

Maxwell, T. and R. Costanza. "Spatial Ecosystem Modeling in a Distributed Computational Environment." In J. van den Bergh and J. van der Straaten, eds., *Toward Sustainable Development: Concepts, Methods, and Policy,* pp. 111-38. Washington D.C.: Island Press, 1994.

_____. "Distributed Modular Spatial Ecosystem Modeling." *International Journal of Computer Simulation: Special Issue on Advanced Simulation Methodologies* 5:3 (1995), 247-62.

_____. "Facilitating High Performance: Collaborative Spatial Modeling." NCGIA. Proceedings Third International Conference/Workshop on Integrating GIS and Environmental Modeling. January 21-25, 1996, Santa Fe, New Mexico.

McKeown, R., D.S. Ojima, T.G.F. Kittel, D.S. Schimel, W.J. Parton, H. Fisher and T. Painter. "Ecosystem Modeling of Spatially Explicit Land Surface Changes for Climate and Global Change Analysis." NCGIA. Proceedings Third International Conference/Workshop on Integrating GIS and Environmental Modeling. January 21-25, 1996. Santa Fe, New Mexico.

Mitchell, K.J. "Simulation of Growth of Even-aged Stands of White Spruce." *Yale University School Forestry Bulletin* 75 (1969) 1-48.

Morain, S. and S. López Baros, eds. *Raster Imagery in Geographic Information Systems.* Santa Fe, New Mexico: OnWord Press, 1996.

Newnham, R.M. "The development of a stand model for Douglas-Fir." Ph.D. diss., University of British Columbia, Vancouver, 1964.

Nuñez Brown, D. and P. Brinckerhoff. "Choosing Efficient, Cost-effective Transportation Routes." In S. Morain and S. López Baros, eds., *Raster Imagery in Geographic Information Systems,* pp. 141-48. Santa Fe, New Mexico: OnWord Press, 1996.

ORNL. "Oak Ridge Natural Laboratory: The Stone Super-Computer." URL: http://www.esd.ornl.gov/facilities/beowulf/, 1998.

Pimm, S.L. "Biodiversity and the Balance of Nature." In E.D. Schulze and H.A. Mooney, eds., *Biodiversity and Ecosystem Function,* pp. 347-59. New York: Springer-Verlag, 1994.

Red Hat. "Red Hat Software: Extreme Linux." URL: http://www.redhat.com/, 1998.

Richardson, G.P. *Introduction to System Dynamics: Modeling with Dynamo.* Cambridge, Massachusetts: MIT Press, 1981.

Ridge, D., D. Becker, P. Merkey and T. Sterling. "Beowulf: Harnessing the Power of Parallelism in a Pile-of-PCs." Proceedings IEEE Aerospace, 1997.

Rohlf, F.J. and D. Davenport. "Simulation of Simple Models of Animal Behavior with a Digital Computer." *Journal of Theoretical Biology* 23 (1969), 400-24.

Schneider, S.H. "Scenarios of global warming." In P.M. Kareiva, J.G. Kingsolver, and R.B. Huey, eds., *Biotic Interactions and Global Change,* pp. 9-23. Sunderland, Massachusetts: Sinauer Associates, Inc., 1993.

Shugart, H.H. and D.C. West. "Development of an Appalachian Deciduous Forest Succession Model and its Application to Assessment of the Impact of the Chestnut Blight." *Journal of Environmental Management* 5 (1977), 161-79.

Shugart, H.H. *Terrestrial Ecosystems in Changing Environments*. Cambridge: Cambridge University Press, 1998.

Slothower, R.L., P.A. Schwarz and K.M. Johnston. "Some Guidelines for Implementing Spatially Explicit, Individual-based Ecological Models within Location-based Raster GIS." NCGIA. Proceedings Third International Conference/Workshop on Integrating GIS and Environmental Modeling. January 21-25, 1996, Santa Fe, New Mexico.

Smith, T.R., J. Su, A. Saran, and A.M. Sastri. "Computational Modeling Systems to Support the Development of Scientific Models." In M.F. Goodchild, L.T. Steyaert, B.O. Parks, C. Johnston, D. Maidment, M. Crane, and S. Glendinning, eds., *GIS and Environmental Modeling: Progress and Research Issues*. Fort Collins, Colorado: GIS World Books, 1996.

Tansley, A.G. "The use and abuse of vegetational concepts and terms." *Ecology* 16 (1935) 284-307.

VSI. "Ventana Systems, Inc." URL: http://www.std.com/vensim/, 1998.

Wesseling, C.G., W.P.A. van Deursen and P.A. Burrough. "A Spatial Modelling Language that Unifies Dynamic Environmental Models and GIS." NCGIA. Proceedings Third International Conference/Workshop on Integrating GIS and Environmental Modeling. January 21-25, 1996, Santa Fe, New Mexico.

Westervelt, J.D. and L.D. Hopkins. "Facilitating Mobile Objects within the Context of Simulated Landscape Processes." NCGIA. Proceedings Third International Conference/Workshop on Integrating GIS and Environmental Modeling. January 21-25, 1996, Santa Fe, New Mexico.

Yuan, M. "Temporal GIS and Spatio-temporal Modeling." NCGIA. Proceedings Third International Conference/Workshop on Integrating GIS and Environmental Modeling. January 21-25, 1996, Santa Fe, New Mexico.

Analyzing Phosphorous Loads and Transport in the Lake Okeechobee Watershed

J. Zhang, W. Guan, and A. Essex, South Florida Water Management District

Resource Management Requirement

Lake Okeechobee in South Florida has experienced increasing eutrophication over the past 25 years. Beginning in the 1970s, the South Florida Water Management District implemented a phosphorus management practices program in the Lake Okeechobee watershed to reduce total phosphorus loads entering the lake (Federico et al. 1981). In 1989, a regulatory program was adopted to limit total phosphorus concentrations in surface water runoff from land parcels in the watershed (SFWMD 1989). Although these two programs have reduced phosphorus loads, annual loads are still more than 93 tons above the legally mandated target (Campbell et al. 1993). To further evaluate the effectiveness of phosphorus management practices on a parcel by parcel basis, a GIS modeling system for analyzing site-specific phosphorus loading and transport (GIS-PLAT) was developed (SFWMD 1997). The system is linked with a relational database management system to identify phosphorus discharge violation sites (referred to here as "noncompliant") throughout the watershed (Zhang and Essex 1997). GIS-PLAT contains a field hydrologic and nutrient transport model (FHANTM) coupled with a GIS to simulate runoff volume and phosphorus loads. It allows water managers to alter land uses and/or phosphorus management practices on site and view resulting phosphorus load reductions. GIS-PLAT stores, retrieves, and presents data; edits modeling scenarios; manages model runs; and generates reports and maps.

This essay describes the GIS-PLAT system design and modeling databases, the specific objectives of which were to (1) develop a framework for GIS-PLAT; (2) prepare flat file databases for temperature, precipitation, soil, land use/ phosphorus management practices, and hydrography; and (3) develop a graphical user interface to display and analyze model output.

Driving Force

The South Florida Water Management District has implemented a phosphorus management practice program in the Taylor Creek/Nubbin Slough and Lower Kissimmee River basins. These basins contribute most phosphorus entering Lake Okeechobee. Chapter 40E-61of the Florida Administrative Code was adopted in 1989 to regulate total phosphorus concentrations in surface water runoff from land parcels larger than 0.2 hectares. Although these programs reduced phosphorus loads in drainages entering the lake, the average from 1990 to 1994 remained higher than the legally mandated target of 397 tons per year. This target was established using a modified Vollenweider model for Florida lakes.

To determine the source and extent of excess phosphorus loads, the authors identified 57 noncompliant sites in the watershed as possible candidates for additional load reductions. Such sites have runoff with average total phosphorus concentrations exceeding legal discharge limits. Based on simulated runoff and measured phosphorus concentrations, noncompliant sites annually contribute an estimated 24 tons of phosphorus. Bringing these sites into compliance could reduce the excess by 17 tons (or 18 percent of the 93-ton excess). Given this stimulus, a need arose for a user-friendly modeling tool that water managers could apply to further view, model, and evaluate both existing and potential load reduction sites.

Data and Methodology

The Lake Okeechobee Agricultural Decision Support System (LOADSS) was designed to allow water managers and planners to view the environmental and economic effects of hypothetical changes in land use and phosphorus management practices on a regional scale (one or many basins) on a long-term, average annual basis (Negahban et al. 1995). However, LOADSS has a predefined database that cannot be modified and was not designed for short-term or site-specific phosphorus loading studies.

GIS-PLAT, which was specifically developed to study non-compliant sites, offers four major capabilities unavailable in LOADSS.

- A method for identifying sites using a relational database management system.

- A method for estimating runoff volumes and phosphorus loads for these sites.

- A method for simulating trends in phosphorus loads under different management strategies through time.

- An ability to add new sites to the analysis.

While LOADSS is used as a screening tool to identify management practices that reduce loads to an acceptable level while minimizing economic impact at a regional scale, GIS-PLAT offers accurate, site-specific load estimates for local water resource managers.

System Design

GIS-PLAT includes a relational database management system for field water quality data, a GIS database for model input and output, a hydrologic/water quality model to analyze data input, and a graphical user interface (see the next illustration). The field hydrologic and nutrient transport model (FHANTM) is the hydrologic/water quality model. The database management system is ORACLE7 running on a VAX mainframe computer. The GIS software used include ARC/INFO version 7 and ArcView version 3 running on

UNIX workstations. Connection between ARC and ORACLE is through SQL*Net and ARC database integrator, customized by ARC macro language (AML) programs. Field hydrologic data and other flat file input are viewed using ARC/INFO or ArcView. A preprocessor was developed using AML to prepare input for FHANTM. After executing FHANTM, a post-processor converts model output (in text format) into a GIS database format. Using the interface programs, all input data are modeled to produce reports, charts, and maps in the GIS.

zhanfig1.tif

Schematic representation of GIS-PLAT components.

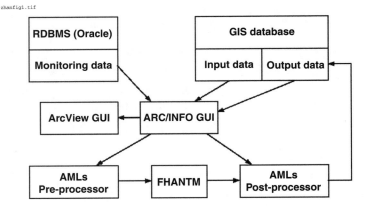

A process/data flow diagram for using FHANTM to generate output for use in GIS-PLAT appears in the following illustration. FHANTM is a field-scale hydrologic/water quality model that simulates flow (surface runoff and subsurface lateral flow) and phosphorus loads per unit area. A *field* is defined as having a single land use, relatively homogenous soils, spatially uniform rainfall, and a single management practice (Knisel 1980). Thus, one or more fields comprise a noncompliant site, and total average annual flow from a site is computed as the sum of flows from each field within the site. A site's areal extent is determined from the drainage area of the closest downstream water quality monitoring station. Field areas were calculated using ARC/INFO.

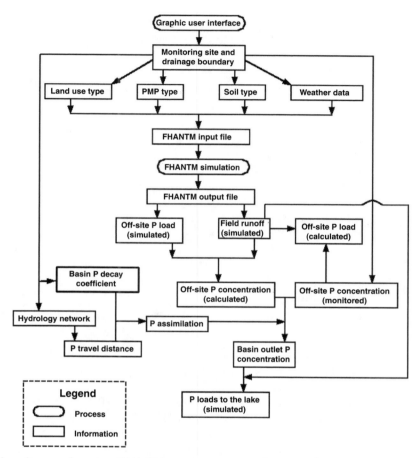

Flow diagram for using FHANTM to generate output for use in the GIS-PLAT system.

FHANTM inputs were divided into three general groups: weather, soil, and model parameter files for each combination of land use and phosphorus management practice. Because most noncompliant sites do not have on-site weather recorders for rainfall and temperature, two weather regions were established: one west and the other east of Kissimmee River (see the next illustration). Rainfall data collected at S-70 and Judson were used for sites located in the west and east regions, respectively. FHANTM parameter files contain generic data that can be used for any noncompliant site. Parameter values can also be modified if site-specific data are available.

zhanfig3.tif

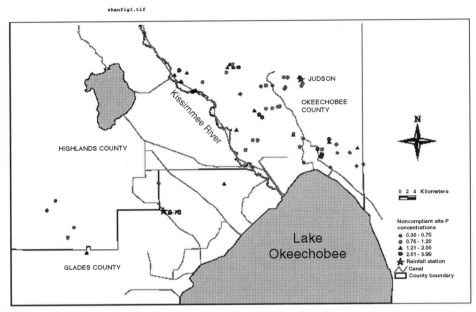

Rainfall stations and noncompliant sites in the Lake Okeechobee watershed.

Average annual phosphorus loads from each site were estimated by adding loads from surface runoff and subsurface lateral flow. To estimate loads under current land use and management scenarios, loads from surface runoff and subsurface lateral flow were obtained from model output. However, if measured phosphorus concentrations were available, loads from surface runoff were computed by multiplying simulated average annual runoff by measured concentration values.

A portion of phosphorus discharged from these sites is retained or assimilated in stream and wetland systems before reaching the lake. Phosphorus retention by the transport system is estimated using the exponential decay function

$$C_o / C_i = e^{-aTL},$$

where C_o is phosphorus concentration at the basin outlet (measured in mg/L), C_i is concentration at the site outlet (or entry into the transport system, mg/L), a is the assimilation coefficient (km^{-1}), and TL is the total length of the transport system (km). Values for a were obtained from the University of Florida. Total stream length was obtained from the hydrography database. Values for C_o/C_i were computed from the equation, and values of C_i for noncompliant sites were used to estimate phosphorus concentrations at the basin outlet.

GIS-PLAT Databases

GIS-PLAT contains FHANTM model input and output databases and a hydrography database. FHANTM model inputs are weather, soil, and model parameter files for each combination of land use and management. Model outputs include runoff, subsurface lateral flow, phosphorus loads in runoff, and phosphorus loads in subsurface lateral flow. Descriptions of model inputs and example data from model output are presented below.

Weather Data

FHANTM requires daily temperature data and hourly rainfall. Because daily temperature data were unavailable at these locations, temperature data were generated using the WGEN program, which is a model for generating daily weather variables (Richardson and Wright 1984). WGEN requires site-specific rainfall data and temperature parameters to generate appropriate data. Temperatures were estimated using measurements collected from Archbold Biological Station, 56km southwest of the Judson station.

Soil Data

Soil data used in the FHANTM simulation were obtained from the University of Florida (1993). These data represent the Smyrna-Immokalee association, the dominant association in the study area. Among the 57 noncompliant sites, the Smyrna-Immokalee soil association comprises 78 percent of the total area.

Phosphorus Management Practices

Parameter files for each phosphorous management practice (PMP) were developed to reflect land use and management scenarios. For noncompliant sites, land uses were categorized into one of seven types: high animal traffic area, heifer pasture, improved beef pasture, ornamental, hay field, row crops, and native range. Management scenarios for these land uses are listed in the following tables. Scenarios listed in the first table show land uses and PMPs for noncompliant sites. Steady state simulations represent typical flow and phosphorus loads assuming the land use and management practice had always been in place for that field. The transient simulations in the second table represent changes from a land use and management practice yielding a high phosphorus discharge to one with a lower discharge.

Land use and PMPs for noncompliant sites based on steady state simulations using FHANTM

Land use	PMP
High animal traffic area	Typical
Heifer pasture	1 heifer/1 acre; 1 heifer/2 acres
Improved beef pasture	1 beef cow/2 acres; 1 beef cow/3 acres; 1 beef cow/4 acres
Hay field	Bahia; Pangola; Callie Bermuda
Row crop (truck crop)	Typical
Ornamental	Typical
Native range	Typical

Land use and PMPs for changing non-dairy land uses

Land use from	Land use to	PMP
High animal traffic area	Heifer pasture; Heifer pasture	1 heifer/1 acre; 1 heifer/2 acres
High animal traffic area	Beef pasture; Beef pasture; Beef pasture	1 beef cow/2 acres; 1 beef cow/3 acres; 1 beef cow/4 acres
High animal traffic area	Hay field; Hay field; Hay field	Bahia; Callie Bermuda; Pangola
High animal traffic area	Row crop	typical
High animal traffic area	Ornamental	Typical
High animal traffic area	Native range	Typical
Heifer pasture (1 heifer/ 1 acre)	Hay field; Hay field; Hay field; Native range	Bahia; Callie Bermuda; Pangola; Typical
Heifer pasture (1 heifer/ 2 acres)	Hay field; Hay field; Hay field; Native range	Bahia; Callie Bermuda; Pangola; Typical
Beef pasture (1 beef cow/2 acres)	Hay field; Hay field; Hay field; Native range	Bahia; Callie Bermuda; Pangola; Typical
Beef pasture (1 beef cow/3 acres)	Hay field; Hay field; Hay field; Native range	Bahia; Callie Bermuda; Pangola; Typical
Beef Pasture (1 beef cow/4 acres)	Hay field; Hay field; Hay field; Native range	Bahia; Callie Bermuda; Pangola; Typical
Hay field: Bahia; Callie Bermuda, Pangola	Native range; Native range; Native range	Typical; Typical; Typical
Row crop	Native range	Typical
Ornamental	Native range	Typical

Hydrography

Phosphorus loads retained by the transport system were related to travel distance between each discharging site and the lake. A hydrography coverage containing travel route (through streams and wetlands) from each discharge site to its basin outlet was developed. Travel length from the outlet of a noncompliant site to the basin outlet was computed using ARC/INFO.

FHANTM Output

FHANTM provides edge-of-field (off-site) estimates of flow and phosphorus loads. Phosphorus loads to the lake are estimated using total flow (adding runoff and subsurface lateral flow) multiplied by the concentrations at the basin outlet. Off-site phosphorus concentrations in total flow, or measured concentrations if available, and the travel length for each site, were used to estimate concentrations at the basin outlet (see the preceding equation).

GIS-PLAT User Interface

The graphical user interface was designed to help managers view spatial data (i.e., land use, hydrography, and soil type), measured water quality data, and model input and output in a user-friendly, menu driven environment (see the next illustration). Noncompliant sites in the Lake Okeechobee watershed can be retrieved and displayed by selecting View Input | OOC View. Model parameters can be viewed by selecting View Input | Model Parameters. Phosphorus loads produced from a site in previous years can be estimated by selecting Edit Scenario | Past and then clicking on the Run Model button to start model execution. Phosphorus loads to be produced from a site under a projected land use and phosphorus management practice condition are simulated by selecting Edit Scenario | Future and then the Run Model button.

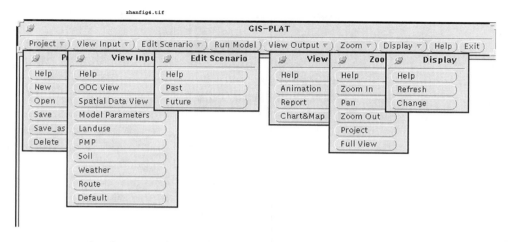

zhanfig4.tif

GIS-PLAT graphical user interface.

Model output can be viewed as reports, charts, or maps, and even in animated form.

A typical GIS-PLAT session incorporates the following steps.

1. Open an existing project/plan or initiate a new project/plan.

2. View noncompliant site distribution and average phosphorus concentration values.

3. Select noncompliant sites for analysis.

4. View data obtained from the ORACLE database and generic model parameter files, and change model parameter values if site-specific data are available.

5. Perform model simulations for existing land use/phosphorus management practice conditions to estimate loads from these sites, or change land use and management practices for selected sites and analyze the effects of these changes on load reductions.

6. Display and examine results through animation, reports, charts, and maps.

Reports, charts, and maps are generated only for sites selected at the time the display option is invoked. This feature allows managers to create reports and maps for site(s) of interest. Examples of charts generated in GIS-PLAT are presented in the following two figures. The first figure shows time series data for surface runoff, lateral flow, and total flow typical of a native range. The second figure below shows phosphorus loading trends that occur after a beef pasture with a stocking rate of one cow per two acres is converted to native range. As expected, phosphorus loading associated with surface runoff, lateral flow, and total flow decrease over time because phosphorus fertilizer applications ceased after the pasture was converted.

Sample graphic output for flow.

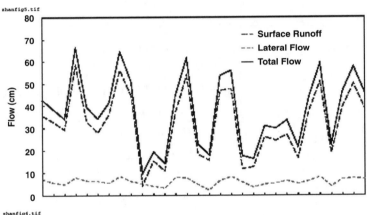

Sample graphic output for P load.

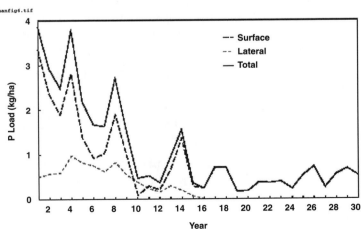

Summary

A framework for a GIS based modeling system to analyze site-specific phosphorus (P) loading and transport (GIS-PLAT) was developed. This tool enables water managers to examine the effects of alternative P management practices for reducing P loads to Lake Okeechobee. GIS-PLAT was linked with a relational database management system to identity noncompliant sites in the Lake Okeechobee watershed. GIS-PLAT also uses a water quality model and a graphic user interface to develop site management plans for reducing P discharge. The system was designed to be as flexible as possible, allowing for additional modifications and application to other watersheds.

Acknowledgments

The authors thank Sharon Wallace and Dera Muszick for providing the GIS expertise and digitizing work, and Gary Ritter and Linda Price for their assistance in completing this project. Appreciation is also extended to Todd Tisdale, Alan Steinman, Gary Ritter, Steffany Gornak, and Joe Albers for their valuable comments and suggestions.

References

Campbell, K.L., A.B. Bottcher, D.A. Graetz, and T.K. Tremwel. "Performance of selected BMPs for phosphorus reduction in high phosphorus source areas." Report submitted to the South Florida Water Management District, West Palm Beach. Gainesville, Florida: Agricultural Engineering and Soil Science Departments, Institute of Food and Agricultural Sciences, University of Florida, 1993.

Federico, A.C., K.G. Dickson, C.R. Kranzer and F.E. Davis. "Lake Okeechobee Water Quality Studies and Eutrophication Assessment." Technical Publication 81-2. West Palm Beach, Florida: South Florida Water Management District, 1981.

Knisel, W.G., ed. "CREAMS: A Field-scale Model for Chemicals, Runoff, and Erosion from Agricultural Management Systems." U.S. Department of Agriculture, Conservation Research Report, No. 26, 640 pp. Washington, D.C.: Department of Agriculture, 1980.

Negahban, B., C. Fonyo, W.G. Boggess, J.W. Jones, K.L. Campbell, G. Kiker, E. Flaig and H. Lal. "LOADSS: A GIS-based Decision Support System for Regional Environmental Planning." *Ecological Engineering* 5 (November 1995), 391-404.

Richardson, C.W. and D.A. Wright. "WGEN: A Model for Generating Daily Weather Variables." Agricultural Research Service, ARS-8, 83 pp. Washington, DC: U.S. Department of Agriculture, 1984.

South Florida Water Management District (SFWMD). "Interim Surface Water Improvement and Management (SWIM) Plan for Lake Okeechobee." West Palm Beach, Florida: South Florida Water Manaement District, 1989.

_____. "Surface Water Improvement and Management (SWIM) Plan Update for Lake Okeechobee." West Palm Beach, Florida: SFWMD, 1997.

University of Florida. "Lake Okeechobee Agricultural Decision Support System: Model Verification/Calibration/Sensitivity Analysis (version 2.2)." Report submitted to the South Florida Water Management District, West Palm Beach. Gainesville: Institute of Food and Agricultural Sciences, Department of Agricultural Engineering and Department of Food and Resources Economics, University of Florida, 1993.

Zhang, J. and A. Essex. "Phosphorus Load Reductions from Out-of-compliance Sites in the Lake Okeechobee Watershed, Florida." *Applied Engineering in Agriculture* 13:2 (1997), 193-98.

Global Commons Risk Assessment

R.C. Lozar and H.E. Balbach,
U.S. Army Construction Engineering Research Laboratories

Resource Management Requirement

At the end of the 1980s, coincident with the decline of the Soviet Union, the West German government requested that the U.S. Army remove its stockpile of chemical filled artillery shells that had been stored there many years earlier as a Cold War deterrent. Under domestic U.S. law, the munitions could be transported only to the Johnston Island chemical agent destruction facility in the central Pacific Ocean. While the West German government oversaw transport from the storage location to the North Sea port of embarkation, the U.S. Army was responsible for evaluating the risks of oceanic transportation. Under Executive Order 12114 and existing regulations for environmental assessment of Federal actions (Army Regulation 200-2 1989 and 200-1 1990), transport through the Global Commons required an environmental impact assessment (EIA). Also applicable, new international standards for shipping such dangerous cargo complicated the physical arrangements for packaging and loading (International Convention 1979). At the time (1990), no technique existed for assessing such a global action, and no existing documents examining global commons issues had ever focused on the oceans and the relative risk of different transportation routes.

The overall goal was to recommend the lowest risk route by sea for the munitions. The environmental assessment study covered movement from the time when the munitions left West German territorial waters to their arrival in U.S. territorial waters. Sea movement of the munitions was accomplished under the International Maritime Dangerous Goods (IMDG) Code (International Convention 1979). Potentially usable routes included the Panama Canal, South Atlantic, and Indian Ocean. From these routes, four primary alternatives based on standard shipping routes for transport of cargo between Western Europe and the mid-Pacific were examined (see the following figure). The approach gauged each route's potential for harming natural resources or coastal populations in the event of an incident.

Four shipping tracks evaluated.

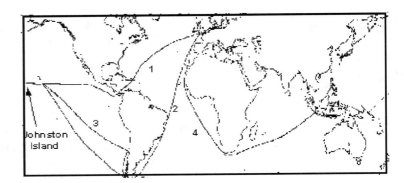

The routes were compared using an extension of the Geographic Resource Analysis Support System (GRASS) (Westervelt et al. 1992). In a special database prepared for the study, 70 global environmental data elements were entered. Each theme consisted of a raster file containing a cell 4'48" in size (0.08 degrees, or about 9km) yielding 10,125,000 data cells per theme. Few databases existed which were relevant to oceanic and coastal areas, and fewer still covered the entire globe. For this study, data sources that contained 125 appropriate environmental topics were identified (Balbach and Lozar 1996).

Resources closest to the proposed track of the ship were at greatest risk. A "trolling theory" algorithm provided a decreasing weighted sum as distance from each route increased. The analysis consisted of seven major areas with 44 topics and 74 subtopics. The final weighting, categorized by subtopic, measured relative sensitivity multiplied by route distance through each zone, multiplied again by importance of the major topic. The database was broadly divided into biological and social (or human) elements. Phytoplankton productivity, marine mammals, sea turtles, and pelagic fisheries are examples of biological factors. Human population density in coastal areas was the primary social element (Riggins et al. 1976, U.S. Army 1987).

The sum of all topics, subtopics, and weightings resulted in the Worldwide Sensitivity Map, which clearly shows the relative sensitivity of coastal areas studied. The next figure displays only basic sensitivity data; no geographic boundaries are presented.

Summary worldwide sensitivity map.

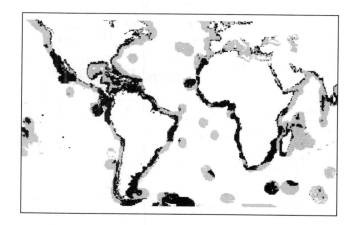

A modified Gaussian plume dispersion model was run to define worst case hazard calculations. The model assigned no likelihood of loss of life distances beyond 3.2km (approximately 2 mi) during the day and 5.7km (approximately 3.5 mi) at night. Because cell size in the database is roughly 9km (5.5 mi), acute human death risk was essentially limited to the cell containing the ship plus, at most, one adjacent cell. The trolling algorithm was modified based on this information, with highest weights being applied to the cell through which the ship was to sail, and slightly lower weights assigned to adjacent cells. This was done for every cell in the database for every environmental factor for each route. The sum of all values represented the "score" for each route.

Results

The EIA showed significantly different total scores among the routes considered. The unit of measure was square miles of shipping track. This was convenient because the latitude/longitude definition of the data cells easily converted to nautical miles. These data were examined in several ways. Longer routes generally showed higher scores (see the next table) because they contained more cells. However, the total scores were not solely proportional to route length. The weighted scores of routes 1 and 4 contained far more cells with hazards above the moderate level, as originally determined by the assigned weights previously mentioned.

Combined scores of four routes for environmental factors and physical hazards (square miles)

Estimated hazard	Route 1	Route 2	Route 3	Route 4
Negligible environmental hazard	6,351	32,243	32,432	22,194
Low hazard	24,6681	5,318	15,995	24,723
Low to moderate hazard	1,648	15,166	16,336	13,712
Moderate hazard	12,441	10,204	16,003	28,566
Moderate to high hazard	4,266	152	115	5,538
High hazard	418	0	0	299
Very high hazard	377	0	0	0
Total sq mi >= low hazard	54,818	40,839	48,449	72,838

Closer examination of each route also revealed other types of differences—as well as some similarities. Prominent among similarities was the revelation that virtually all high and very high scores were related to potential for hazard to humans, which were in turn associated with places where the ship passed within 5km of a populated coastal area. Several zones of moderate to high hazard were associated with known coastal sensitive areas other than those associated with human populations (e.g., turtle nesting beaches, shore bird sanctuaries, and high-value coastal wetlands). However, in all cases the common thread was close passage to a coastal area. The map of relative environmental sensitivity (see the previous figure) displays this information, although its importance was undetected until after the results were examined. The next table presents the proportions of each route greater than or equal to a moderate hazard level.

Proportion of route rated moderate hazard or higher

Route	1	2	3	4
Percentage	28	13	19	37

To further reduce risks to both sensitive natural and social environments, routes were modified to decrease the length of time adjacent to land. Although this decreased the absolute scores, the relative rankings of the alternative routes remained unchanged despite these modifications. The ultimate recommendation was to discard routes which came within 10km of any land mass if an alternative was available.

This case study, which represents the first implementation of a "global commons EIA," was apparently highly successful because mission planners ultimately used the input to determine the final route chosen—route 2—even with other information available to them. While route 2 was not the shortest, it did display the fewest high-risk elements.

The documentation associated with the Global Commons EIA withstood two Federal court challenges (Greenpeace USA V. Stone. 748 F. Supp. 749 District Court of Hawaii 1990 and its appeal in the Ninth Circuit Court of Appeals in San Francisco, September 1990). In the latter, the Government successfully argued that the proposed ocean transport should not be halted, and that potential accidental effects on ocean bottom life had been adequately considered in the EIA (Balbach et al. 1990). The court opinion concluded that although foreign policy implications must be taken into consideration, EO 12114 did not "preempt application of NEPA [National Environmental Policy Act] to all federal actions taken outside the United States." (The transfer was carried out without incident in September-December 1990.) Thus, this GIS based work entered case law as one of the first legal tests of resource management in a global context (Balbach and Lozar 1996).

Conclusion

U.S. laws, regulations, and international treaty obligations can in part be supported through innovative application of geographic information systems. The relatively small number of data sources available for the global commons study at the time it was conducted have since been greatly expanded, and the concept of a "global database" is now firmly entrenched. Integrating data and decisions into an EIA lends credibility to decisions and is useful in withstanding court challenges.

Disclaimer

This essay is not an official publication of the U.S. Army or Department of Defense. The opinions expressed are the authors', and do not represent official findings or policy of the U.S. government.

References

Army Regulation 200-1. "Environmental Protection and Enhancement," 23 April 1990.

Army Regulation 200-2. "Environmental Effects of Army Actions," 23 January 1989.

Balbach, H.E., R. Lacey, M. Chawla, and R.C. Lozar. "Global Commons Environmental Assessment." U.S. Army Office of the Assistant Secretary for Chemical Demilitarization, 1990.

Balbach, H.E. and R.C. Lozar. "Modeling the Global Environment: The Global Commons Environmental Assessment." In Proceedings of Advances in Scientific Computing and Modeling, edited by S.K. Dey. Charleston, Illinois: Eastern Illinois University, May 1996.

Butts, Kent H. "Environmental Security: What is DoD's Role?" Strategic Studies Institute/U.S. Army War College, Carlisle Barracks, Pennsylvania, May 28, 1993.

International Convention for the Safety of Life at Sea. Regulation 2, Part A, Chapter VII, "International Maritime Dangerous Goods Code," 37 UST 47, 1979.

Jain, R.K., T.A. Lewis, L.V. Urban, and H.E. Balbach. "Environmental Impact Assessment Study for Army Military Programs." CERL Interim Report D-13. Champaign, Illinois, December 1973.

National Environmental Policy Act. PL 91-190, 42 U.S.C.A. 4321 to 4370c.

Riggins, R. and E. Novak. "Computer-Aided Environmental Impact Analysis for Mission Change, Operations and Maintenance, and Training Activities: User Manual." CERL Technical Report E-85. Champaign, Illinois, February 1976.

U.S. Department of the Army. "Personal Computer Program for Chemical Hazard Prediction (D2PC)." Chemical Research, Development and Engineering Research Center Report CRDEC-TR-87021, 1987.

U.S. Department of Defense. DODD 6050.67, DODD Reg. 7220.1 and AR 37U41.

U.S. District Court of Hawaii 748 F. Supp. 749, Greenpeace USA V. Stone, 1990.

U.S. Executive Order 12114, 04 January 1979, "Environmental Effects Abroad of Major Federal Actions."

Westervelt, J.A., M. Shapiro, W. Goran, and D. Gerdes. "Geographic Resources Analysis Support System (GRASS), Version 4.0 User's Reference Manual, Revised. U.S. Army Construction Engineering Research Laboratory, ADP Report 87-22 (rev), 1992. (Available through National Technical Information Service.)

Planning Emergency Response at a Federal Laboratory

R.Greene, Los Alamos National Laboratory

Resource Management Requirement

The Los Alamos National Laboratory (LANL) occupies approximately 110 sq km on the Pajarito Plateau in northern New Mexico. The LANL complex includes 50 technical sites scattered throughout an area with rugged terrain and complex meteorology. Many of the more than 2,000 structures in the LANL complex house hazardous materials—ranging from the most plentiful, chlorine, to the most recognized, plutonium. LANL employs nearly 13,000 people, many of whom live in the nearby communities of Los Alamos and White Rock. Others commute from the adjacent counties of Santa Fe and Rio Arriba. The combined population of the three counties is nearly 170,000.

Most atmospheric release scenarios produce no significant risk beyond 2km. However, specific combinations of wind conditions, a particular hazardous material, and its release rate can affect areas up to 10km away. Depending on wind direction, this scale of release can affect the Los Alamos and White Rock communities as well as the LANL complex. The terrain of the Pajarito Plateau features many steep-sided mesas and canyons, and there are few evacuation routes.

The Emergency Management and Response (EM&R) Team at LANL uses GIS technology to augment their system for mitigating accidental atmospheric releases of toxic and hazardous materials. GIS affords fast visual analysis of diverse data layers, including atmospheric conditions, atmospheric plumes, population centers, transportation routes, facilities, and terrain.

Methodology

EM&R currently uses the Meteorological Information and Dose Assessment System (MIDAS) software package to predict current and projected isopleths for radiological and chemical hazard levels. These plume calculations are performed in near real time and are based on current estimates of total release and release rate, as well as the current wind field snapshot. The wind field is updated every 15 minutes from four meteorological observation towers located around LANL (Stone and Holt 1995).

The MIDAS modeling tool has radiological (R-MIDAS) and chemical (C-MIDAS) modules. Both models are based on the concept of a segmented plume. The amount of material in each plume segment is determined by the rate at which material is released. The growth rate or dispersion of plume segments is modeled in a wind field interpolated from current measurements of speed direction and fluctuation. These measurements are automatically updated via the tower network. MIDAS output consists of a text file with XY coordinates of the outlines of current and projected plumes. C-MIDAS calculates plume outlines for areas affected by concentrations exceeding Emergency Response Planning Guide levels 2 and 3. R-MIDAS calculates isopleths of total effective dose equivalent and projected dose using EPA-400 guidance.

EM&R uses ArcView for analysis and display of geographic data (ESRI 1995). All geographic data are stored in a format compatible with ArcView. However, the MIDAS output must be treated differently for two reasons. Because the plume outlines are output as text files, they must be converted into ArcView (shape file) format. Next, because these outlines are continually changing, a method is needed to allow retrieval of the latest information for display.

The conversion of plume outlines into shape file format is implemented with the help of the Avenue programming language, an object-oriented language incorporated in Arc-View. Avenue code was developed to read the text file output by MIDAS and create a polygon coverage containing the plume outline and other information in the form of attributes. The shape files are added to the project's main view as themes, which can be turned on or off at the user's discretion. The project's main view also has many other built-in themes, and the user may turn these other themes on or off in order to answer questions specific to a given emergency.

A scenario of how ArcView can display affected areas by a hypothetical plutonium release (main view). A listing of buildings in the evacuation zone has been produced using tools within ArcView.

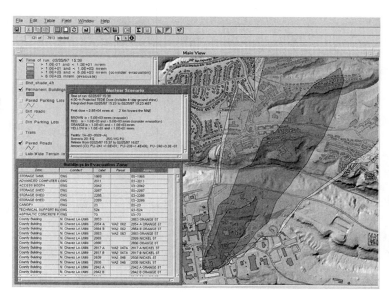

Because a text file is used to send information from MIDAS to ArcView, it was necessary to avoid the problem of two processes accessing the same file at the same time. This was accomplished through the use of semaphores, or flag files. The existence or nonexistence of semaphores serves as a signal between the two software applications, allowing them to take and then release control of the text file asynchronously. Consequently, the problem of concurrent access is avoided.

Results

Timely evacuation of threatened areas requires an up-to-date display of the plume information in relation to facilities, residences, and transportation routes. GIS software has facilitated the integration of data from many diverse sources. ArcView provides many of the classic GIS analytical operations with no modification. Furthermore, the Avenue programming language provides the ability to add specific functionality, such as the data file conversion from the MIDAS data feed. Other operations can be streamlined, such as generating a list of transportation routes or buildings in an affected area. This could be combined with other applications for broadcasting warnings and advisory information.

Managers at the LANL Emergency Operations Center (EOC) can view current and projected areas at risk and therefore make informed decisions regarding necessary protective actions for mitigating an atmospheric release. This integrated application also allows advance emergency planning for specific release scenarios.

References

Environmental Systems Research Institute. "ArcView Functional Overview." ArcView White Paper Series. Redlands, California: ESRI, February 1995.

Stone, G. and D. Holt. "Meteorological Monitoring at Los Alamos." LA-UR-95-3697. Los Alamos: LANL, 1995.

Contributors

Harold Balbach
CERL-LLP
P.O. Box 9005
Champaign, IL 61826-9005
h-balbach@cecer.army.mil
217.373.4560 or 800.USA.CERL, ext 4560

John G. Bartlett
5755 Nutting Hall
University of Maine
Department of Wildlife Ecology
Orono, ME 04469-5755
jbartlett@apollo.umenfa.maine.edu

Teri Brotman Bennett
Earth Data Analysis Center
Bandelier West Room 111
University of New Mexico
Albuquerque, NM 87131-6039
505.277.3622
tbennett@spock.unm.edu

Ward Brady
College of Architecture and Environmental Design
Environmental Resources Program
Arizona State University
Tempe, AZ 87108-2005
602.965.2402
ward.brady@asu.edu

Amy Budge
Earth Data Analysis Center
Bandelier West Room 111
University of New Mexico
Albuquerque, NM 87131-6039
505.277.3622, x231
505.277.3614 fax
abudge@spock.unm.edu

Russell G. Congalton
Department of Natural Resources
215 James Hall
University of New Hampshire
Durham, NH 03824-3589
603.862.4644
603.862.4976 fax
russ.congalton@unh.edu

David J. Cowen
Liberal Arts Computing Lab
University of South Carolina
Columbia, SC 29208
803.777.6803 phone
803.777.7489 fax
cowend@sc.edu

Bruce Dahlman
System Administration
Department of Natural Resources
Information Technology Division
P.O. Box 47020
Olympia, WA 98504-7020
360.902.1055
bruce.dahlman@dnr.state.wa.us

Stephen Egbert
Department of Geography and
Kansas Applied Remote Sensing Program (KARS)
University of Kansas
Lawrence, KS 66045
913.864.7719 or 7708
913.864.7789 fax
s-egbert@ukans.edu

Barbara Entwisle
Department of Sociology
University of North Carolina
Chapel Hill, NC 27516-3997
919.966.1713
919.966.6638, fax
entwisle@unc.edu

Annoesjka Essex
Okeechobee Service Center
South Florida Water Management District
Okeechobee, FL 34973
941.462.5260
aessex@sfwmd.gov

Joyce M. Francis
Department of Plant Biology
Arizona State University
Tempe, AZ 85287-1601
602.965.4685

Kass Green
Pacific Meridian Resources
5915 Hollis Street
Bldg. 8
Emeryville, CA 94608
510.654.6980

Robert Greene
Los Alamos National Laboratory
Mail Stop D452
Warehouse SM30
Bikini Rd.
Los Alamos, NM 87545
505.667.5061
greene@lanl.gov

Robert M. Greenwald
HSI GeoTrans
46050 Manekin Plaza
Ste. 100
Sterling, VA 20166
703.444.7000
703.444.1685 fax

Weihe Guan
Department of Ecosystem Restoration
South Florida Water Management District
3301 Gun Club Rd.
West Palm Beach, FL 33406
561.682.6687
wguan@sfwmd.gov

Stephen Hodge
Florida State University
C2200 University Center
Florida Resources and Environmental Analysis Center
Tallahassee, FL 32306-2641
850.644.2882
shodge@opus.freac.fsu.edu

Steven Holloway
Department of Geography
University of Montana
Missoula, MT 59812
406.542.0535 or 406.728.9546
oikos@bigsky.net

Jeffrey M. Klopatek
Department of Plant Biology
Arizona State University
Tempe, AZ 85287-1601
602.965.4685
klopatek@asu.edu

Bob Lozar
CERL-LL
P.O. Box 9005
Champaign, IL 61826-9005
217.398.5390 or 800.USA.CERL, ext. 5390
r-lozar@cecer.army.mil or lozar@diego.cecer.army.mil
http://www.cecer.army.mil/ll/WWWDEMO/global_apps/
global_menu.html

Deirdre Mageean
Coburn Hall 15
University of Maine
Smith Center for Public Policy
Orono, ME 04469
207.581.2799
Deirdre_Mageean@umit.maine.edu

Kurt Menke
Bandelier West Room 111
University of New Mexico
Albuquerque, NM 87131-6039
505.277.3622, x227
kmenke@unm.edu

Leslie Moore
Department of Ecosystem Restoration
South Florida Water Management District
3301 Gun Club Rd.
West Palm Beach, FL 33406
561.682.6610
lmoore@sfwmd.gov

Stan Morain, editor
Earth Data Analysis Center
Bandelier West Room 111
University of New Mexico
Albuquerque, NM 87131-6039
505.277.3622, x228
505.277.3614 fax
smorain@spock.unm.edu

Paul R.H. Neville
Bandelier West Room 111
University of New Mexico
Albuquerque, NM 87131-6039
505.277.3622
pneville@spock.unm.edu

Raymond O'Connor
238 Nutting Hall
Department of Wildlife Ecology
University of Maine
Orono, ME 04469-5755
207.581.2880
207.581.2858 fax
oconnor@umenfa.maine.edu

Kevin P. Price
Department of Geography and
Kansas Applied Remote Sensing Program (KARS)
University of Kansas
Lawrence, KS 66045
k-price@ukans.edu

Linda Price
Okeechobee Service Center
South Florida Water Management District
Okeechobee, FL 34973
941.462.5260
lprice@sfwmd.gov

Stephen Prince
Department of Geography
University of Maryland
Computer Space Science Bldg., Rm. 2309
College Park, MD 20742
301.405.4050
301.314.9299 fax
sp43@umail.umd.edu

Roland L. Redmond
Wildlife Spatial Analysis Lab
Montana Cooperative Wildlife Research Unit
University of Montana
Missoula, MT 59812
406.243.4906
406.243.6064 fax

Ronald R. Rindfuss
Department of Sociology
University of North Carolina
Chapel Hill, NC 27516-3997
919.966.7779
919.966.6638, fax
rindfuss.cpc@mhs.unc.edu

Gary Ritter
Okeechobee Service Center
South Florida Water Management District
Okeechobee, FL 34973
941.462.5260
gritter@sfwmd.gov

Victor Sánchez-Cordero
Instituto de Biología
Universidad Nacional Autónoma de México
D.F. 04510, Mexico
victor@dunsun.dti.uaem.mx

James Schumacher
Wildlife Spatial Analysis Lab
Montana Cooperative Wildlife Research Unit
University of Montana
Missoula, MT 59812
406.243.4906
406.243.6064 fax

Keith R. Smith
Environmental Health Section Director
Thurston County Public Health Department
2000 Lakeridge Drive SW
Olympia, WA 98502-6045
360.786.5456
360.754.4462, fax
smithk@co.thurston.wa.us

Bill Solecki
350 Mallory Hall
Department of Earth and Environmental Studies
Montclair State University
Upper Montclair, NJ 07043
973.655.5129
973.655.7047 fax
soleckiw@saturn.montclair.edu

P. Srinivasan
HSI GeoTrans
46050 Manekin Plaza
Ste. 100
Sterling, VA 20166
703.444.7000
Fax 703.444.1685
srini@hsigeotrans.com

A. Townsend Peterson
Museum of Natural History
University of Kansas
Lawrence, KS 66045
mexbidiv@falcon.cc.ukans.edu

Robert Walker
Resources for the Future
1616 P St. NW
Washington, DC 20036

Steve J. Walsh
Department of Geography
University of North Carolina
Chapel Hill, NC 27599-3220
919.962.3867
919.962.1537, fax
walsh@geog.unc.edu

David S. Ward
46050 Manekin Plaza
Ste. 100
Sterling, VA 20166
703.444.7000
703.444.1685 fax
davew@geotrans.com

Gary L. Whysong
College of Architecture and Environmental Design
Environmental Resources Program
Arizona State University
Tempe, AZ 87108-2005
gary.whysong@asu.edu

Alain Winckell
El Colegio de la Frontera Norte/ORSTOM
dhenders@rohan.sdsu.edu

Richard Wright
San Diego State University
5500 Campanile Drive
San Diego, CA 92182-4493
619.594.5466, x4597
619.594.4938 fax
wright@typhoon.sdsu.edu

Jiansheng Yan
South Florida Water Management District
Water Use Division
3301 Gun Club Rd.
South Palm Beach, FL 33406
561.686.8800
jason.yan@sfwmd.gov

Joyce Zhang
Department of Ecosystem Restoration
South Florida Water Management District
3301 Gun Club Rd.
West Palm Beach, FL 33406
561.682.6341
jzhang@sfwmd.gov

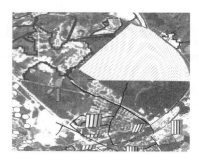

Index

A

accuracy
   of maps 131–137
   of models 13
air pollution 75
algorithms, for models 12
analytical models 303
aquatic resource management
   habitat mapping 121–122
   ownership exhibits, creating 116
   shellfish bed mapping 118–119
   weed control 124
attribute data, in transnational GIS 80

B

base cartographic data. *See* framework data
biodiversity preservation 141–150
business plan, for framework infrastructure 68–69

C

cadastral, in framework infrastructure 55
CAMS data (Calibrated Airborne Multispectral Scanner data) 222–228
CART (classification and regression tree) analysis 200
census tracts 90
choropleth maps 284–285, 288–289
communications, transnational 74
composition metrics 167–169
computer skills and resources, and dynamic models 307–308, 314–316
connectivity indices 171–172
contagion 171
Continental Divide National Scenic Trail (CDNST). *See* trail routing
Continental Divide Trail Alliance 108

Continental Divide Trail Society 108
curtain wall 95, 102

D

dasymetric mapping
   defined 285
   of population density 285–289
data exploration 12
data mining. *See* quantitative analysis
definitive models 10
demographic data. *See* population density
DEMs (digital elevation models) 130, 135
descriptive models 6–8, 300
deterministic models 302
DGPS (differentially corrected global positioning system) 118–119
diversity indices 170–171
dominance indices 170–171
DTMs. *See* DEMs (digital elevation models)
dynamic models
   advances in 306–308
   computer resources for 314–316
   construction of 308–314
   described 302, 305–306
   Internet resources for 316–317

E

ecosystems
   defined 1
   human impact on 2
   management of 268–280, 294–295
ecotopes, defined 169
EIA (environmental impact assessment). *See* environmental risk assessment
elevation data
   and map scale
   harmonizing in transnational GIS 77

in framework infrastructure 53
emergency response planning 347–350
empirical models 300
environmental activism 9
environmental degradation, in transnational
        GIS 76
environmental risk assessment
   in Global Commons 339–344
   in transnational GIS 75
environment-population interactions 193–204,
        251–264
error, map
   measuring 135–137
   types 131–135
expert opinion 13
explanatory models 301
extent, defined 21
external variables 296

F

famine early warning 235–249
feature identification code 60–62
feature, defined 60
FGDC (Federal Geographic Data
        Committee) 45–48
field work, benefit to mapping 119, 121
forcing functions 296
forest economics 7–8
fractal dimension 175
fragmentation 172–173, 279
framework data
   for trail routing 112–114
   in transnational GIS 76–80
framework infrastructure
   administrative issues in 63–66
   business plan for 68–69
   cooperation in 50–51, 66–67
   data themes in 49, 52–55
   development of 48–51
   technical issues in 56–62

G

geodetic control, in framework infrastructure 52
geographic information systems. *See* GIS
geology theme, in transnational GIS 87
geomorphology theme, in transnational GIS 87–
        88
GIS
   building models with 12–13

defined 5
focus of 14
specifying models with 10–11
GIS-PLAT
   databases in 331–334
   design of 327–331
   graphical user interface 334–336
   objectives 326
   vs. LOADSS 327
gliding box analysis 27–34
governmental units, in framework
        infrastructure 55
grain size, defined 21
ground water modeling systems 95–104
GWZOOM 95–104

H

habitat mapping
   aquatic 121–122
   in transnational GIS 75
hazardous materials release, response to 347–
       350
historical data, in framework infrastructure 62
horizontal data integration 57, 81–91
human behavior
   and land use/land cover 251–264
   and primary production 183–190, 243–245
hydrographic data
   in framework infrastructure 54
   in phosphorus transport model 154–161
   in transnational GIS 78–79
hypsographic data, in transnational GIS 76–77

I

image processing, vs. vector based GIS 212–213
infrastructures, and transnational GIS 73, 75
Internet resources, for dynamic models 316–317

J

join-count analysis 22–26

L

lacunarity analysis 26–34
Lake Okeechobee watershed, described 152–154
land cover classifications 167–169
land demand, defined 276
land ownership data
   and trail routing 109, 111–113
   in aquatic resource management 116

land supply, defined 276
land use
    and human behavior 251–264
    and trail routing 109
    change in 275–280
    in transnational GIS 89
landform, defined 88
landscape ecology 18
landscape, defined 87

M

maps
    errors in 131–137
    historical use of 8
    scale of 129–131
metadata 58–60
metric independence 177–178
models
    building 12–13
    characteristics of good 297–299
    defined 296
    types 6–8, 299–304
models. *See also* dynamic models
Monte Carlo simulations 23

N

National Trails System Act 107
natural resource depletion 73
nearest-neighbor analysis 22
NSDI (National Spatial Data Infrastructure) 45, 47

O

orthoimagery 52

P

patch cohesion 172
patches, defined 169
pattern metrics 166, 169, 260, 262
phosphorus transport model
    described 151–152
    hydrograpic data for 154–161
    See also GIS-PLAT
    uses of 161
point data analysis
    techniques for 22–34
    vs. patch data analysis 20–21
population density
    and global vegetation 188–189

dasymetric mapping of 285–289
population growth, and land use change 276
population surveys, and remote sensing
        data 252, 255–257
population-environment interactions 193–204,
        251–264
predictive models 6–8, 11
prescriptive models 6–8, 11
primary production
    human impact on 183–190
    human response to variations in 243–245
    measuring 242–243
profile curvature 260

Q

quantitative analysis 12–13

R

reach
    defined 54
    numbering 157
remote sensing
    and population surveys 252, 255–257
    computing environments for 211–213
    for landscape characterization 252
    for primary production measurement 242–243
    spatial resolution of 210–211, 214–228
resource data, in transnational GIS 81–91
roads
    and trail routing 113
    in transnational GIS 79–80

S

Sahel, described 238–241
scale
    defined 20
    problems with 129–131, 166, 175–177
sewage infrastructures 75
shape metrics 173–175
shellfish bed mapping 118–119
silviculture 7
simulation models 303
slope curvature 260
soil taxonomy 84–85
soils theme, in transnational GIS 83–86
spatial autocorrelation 136
spatial data
    errors in 131–137

for trail routing 112–114
in dynamic modeling 304–316
in transnational GIS 71–91
scale of 129–131
spatial resolution, for remote sensing 210–211,
    214–228
spatially explicit models 304
species distribution modeling 141–150
SPOT data 214–222
state variables 296
static models 301–302
statistics, first-order vs. second-order 19
Stern, Paul 193
stochastic models 302–303
Strahler stream order 78
sustained yield 7
SVI (spectral vegetation index) 184–185

T

temporal change 176–177, 302, 304
Tijuana River watershed GIS. *See* transnational
    GIS
Tijuana River watershed, described 72
Topographic Convergence Index 260
topographic curvature 260
topography, and trail routing 112–113
    See also DEMs (digital elevation models)
toxic material release, response to 347–350
trail routing
    and land ownership data 109, 111–113
    and topography 112–113

proposed GIS/GPS use for 105
spatial database for 112–114
transnational GIS
    described 71–72
    geospatial data integration in 76–91
    issues in 73–76
transportation data, in framework
    infrastructure 54

V

validation 298
vegetation production. *See* primary production
vegetation theme, in transnational GIS 81–82
verification 298
vertical data integration 58

W

water infrastructures 75
water pollution 73, 75
water quality monitoring. *See* phosphorus
    transport model
water runoff modeling, in transnational GIS 75
water sources, and trail mapping 114
watershed boundaries, and trail routing 113
watershed management, transnational 71–91
Web resources, for dynamic models 316–317
weed control 124
wetland fragmentation 279
Wetness Index 260
Wolf, Jim 108, 114

More OnWord Press Titles

NOTE: All prices are subject to change.

Geographic Information Systems (GIS)

GIS: A Visual Approach
$42.95

The GIS Book, 4E
$39.95

*Focus on GIS Component Software
Featuring ESRI's MapObjects*
$52.95 Includes CD-ROM

GIS Data Conversion
$49.95

*GIS Online: Information Retrieval, Mapping,
and the Internet*
$46.95

INSIDE Autodesk World
$49.95 Includes CD-ROM

INSIDE MapInfo Professional
$54.95 Includes CD-ROM

Minding Your Business with MapInfo
$39.95

MapBasic Developer's Guide
$52.95 Includes Disk

*Raster Imagery in Geographic Information
Systems* Includes color inserts
$59.95

INSIDE ArcView GIS, 2E
$46.95 Includes CD-ROM

ArcView GIS Exercise Book, 2E
$49.95 Includes CD-ROM

ArcView GIS/Avenue Developer's Guide, 2E
$52.95 Includes Disk

*ArcView GIS/Avenue Programmer's
Reference, 2E*
$52.95

ArcView GIS /Avenue Scripts: The Disk, 2E
Disk $99.00

ARC/INFO Quick Reference
$26.95

INSIDE ARC/INFO, Revised Edition
$59.95 Includes CD-ROM

*Exploring Spatial Analysis in Geographic
Information Systems*
$52.95

*Processing Digital Images in GIS:
A Tutorial for ArcView and ARC/INFO*
$52.95 Includes CD-ROM

OnWord Press

OnWord Press books are available worldwide from OnWord Press and your local bookseller. For order information, terms, or listings of local booksellers carrying OnWord Press books, call toll-free 1-800-4-ONWORD (1-800-466-9673) or 505-474-5130; fax 505-474-5030; write to OnWord Press, 2530 Camino Entrada, Santa Fe, New Mexico 87505-4835, USA, or e-mail orders@hmp.com. OnWord Press is a division of High Mountain Press.

Comments and Corrections

Your comments can help us make better products. If you find an error, or have a comment or a query for the authors, please write to us at the address below, send e-mail to cleyba@hmp.com, or call us at 1-800-223-6397.

OnWord Press, 2530 Camino Entrada, Santa Fe, NM 87505-4835 USA

Visit us on the Web at http://www.onword.com